Lecture Notes in Applied and Computational Mechanics

Volume 87

Series editors

Peter Wriggers, Leibniz Universität Hannover, Hannover, Germany
e-mail: wriggers@ikm.uni-hannover.de
Peter Eberhard, University of Stuttgart, Stuttgart, Germany
e-mail: peter.eberhard@itm.uni-stuttgart.de

This series aims to report new developments in applied and computational mechanics—quickly, informally and at a high level. This includes the fields of fluid, solid and structural mechanics, dynamics and control, and related disciplines. The applied methods can be of analytical, numerical and computational nature.

More information about this series at http://www.springer.com/series/4623

Ioannis Vardoulakis

Cosserat Continuum Mechanics

With Applications to Granular Media

Springer

Ioannis Vardoulakis (Deceased)
National Technical University of Athens
Athens, Greece

English Edition by Eleni Gerolymatou, Jean Sulem, Ioannis Stefanou and Manolis Veveakis

ISSN 1613-7736 ISSN 1860-0816 (electronic)
Lecture Notes in Applied and Computational Mechanics
ISBN 978-3-030-06986-5 ISBN 978-3-319-95156-0 (eBook)
https://doi.org/10.1007/978-3-319-95156-0

Printed on acid-free paper

This Springer imprint is published by the registered company Springer International Publishing AG part of Springer Nature
The registered company address is: Gewerbestrasse 11, 6330 Cham, Switzerland

Foreword

The original work *Théorie des corps déformables* of the Cosserat brothers (Eugène who was a mathematician and François who was an engineer) was published in 1909. It was based on differential geometry theory applied to mechanics. Extending the concepts of Cauchy on continuum mechanics, the Cosserat brothers developed a theory for continuous oriented bodies that consist not of material points, but also of directions associated with each material point. They recognized the application of their theory for representing the deformations of rods and shells; however, their work was ignored for half a century. New interest in Cosserat continuum theory arose with the rebirth of micromechanics in the 1960s. Different names have been given to Cosserat theory (e.g. micropolar media, oriented media, continuum theories with directors, multipolar continua, microstructured or micromorphic continua). The state of the art at this time was reflected in the collection of papers presented at the historical IUTAM Symposium on the "Mechanics of Generalized Continua", in Freudenstadt and Stuttgart in 1967 (E. Kröner, ed, Springer-Verlag, Berlin, 1968). There is no doubt that Cosserat continuum theory is mostly suitable for describing the kinematics of granular media; this was clear in the minds of the scientists of this first period, among whom Mindlin (*Micro-structure in linear elasticity*. Arch. Rat. Mech. Anal, 10, 51–77, 1964) is the most prominent proponent. However, early applications of Cosserat theory for the description of the mechanics of granular media were less encouraging, as it appeared that Cosserat effects are negligible when the dominant wavelength of the deformation field is large as compared to the grain size.

New interest appeared in the 1980s when the link was made by H. B. Mühlhaus and I. Vardoulakis between Cosserat continuum description and strain localization in their seminal paper: *The thickness of shear bands in granular materials* (Géotechnique, 37(3):271–283, 1987). Later, advanced experimental testing and discrete element model simulations evidenced significant grain rotations inside the shear band.

In the last 30 years, an important literature was published on Cosserat continua with applications to geotechnics and geomechanics (e.g. borehole stability, soil–structure interaction, layered and blocky rock mass, slope stability), structural geology and geophysics (e.g. mechanics of folding and faulting, fault mechanics), structural mechanics (e.g. masonry structures) and more generally in applied mechanics for representing the behaviour of heterogeneous and periodic materials or structures. These developments were associated with studies on constitutive modelling and advanced numerical models among which I. Vardoulakis and co-workers made significant and innovative contributions. I. Vardoulakis also brought clarifications in highly debated issues on the applicability of Cosserat continuum theory to granular materials with respect to asymmetry of stress in granular media, upscaling methods for defining stresses and couple stresses from contact forces, micromechanical interpretation of stresses and couple stresses.

Despite the fact that Cosserat continuum models are becoming more and more popular for applications in various fields of mechanics, the basic concepts of the theory are rarely taught in graduate schools. Maybe this is due to the lack of a comprehensive textbook clarifying the basic concepts of Cosserat continuum theory. In 2009, Prof. Vardoulakis started to write this work on Cosserat continuum mechanics and mechanics of granular media, emphasizing its sound mathematical formulation based on continuum thermodynamics. An original use of the von Mises motor mechanics was introduced, for the compact mathematical description of the mechanics and statics of Cosserat continua. This book contains numerous examples and exercises and addresses postgraduate students and researchers. I. Vardoulakis intended to teach these topics in several advanced graduate and doctoral programmes over the world. The first part of the textbook was almost complete in September 2009 when Prof. Vardoulakis passed away in a tragic accident. With the approval of his family, the book was prepared for publication in Springer series "Lecture Notes in Applied and Computational Mechanics". We believe that it will be of great use for scientists and engineers for addressing advanced multi-scales problems in mechanics.

Paris, France Jean Sulem

Contents

Chapter 1
Introduction

Abstract This chapter succinctly describes the need for a compact representation in order to describe continua with higher degrees of freedom than the classical translational ones.

There is a continuing discussion concerning the "origins" of the so-called advanced-continuum theories, such as the Cosserat Theory. A footnote in a paper by Mindlin reveals the learned opinion on the subject by M. A. Biot, who gives the credit to Cauchy.[1] At any rate as precursors of the Cosserat theory are mentioned in the literature the theory of Lord Kelvin concerning the light-aether and the works of W. Voight on the physics of crystalic matter [1, 2]. A historical note on the subject can be found in the introduction of the CISM Lecture on "Polar Continua" by Stojanović [3].

What is the major difference between classical continua and Cosserat continua? Classical continuum mechanics is based on the axiom that the stress tensor is symmetric. According to Schaefer [4], it is Hamel [5] who has named this statement the *Boltzmann axiom*, since it is Boltzmann who has pointed first, already in the year 1899, to the fact that the assumption about the symmetry of the stress tensor has an axiomatic character. Thus, the Mechanics of continua with non-symmetric stress tensor may be termed also as non-Boltzmann Continuum Mechanics. Such a theory is the theory of the Cosserat continuum, that originates from the seminal work of the brothers Eugène and François Cosserat [6]. This work is a difficult reading that was made known to the general continuum mechanics community through the works of Sudria [7], published in 1935, and through the famous 1958 paper of late Professor Günther [8], who presented the subject using modern tensor notation.

A 3D Boltzmann continuum is a continuous manifold of material *points* that possess 3 degrees of freedom (dofs), those of displacement. The Boltzmann continuum is juxtaposed to the Cosserat continuum, that is in turn a manifold of oriented *rigid particles*, called "trièdres rigides" or *rigid crosses*, with 6 dofs,

[1]Cauchy, A. L. (1851). Note sur l'équilibre et les mouvements vibratoires des corps solides. Comptes-Rendus, 32, 323–326.

I. Vardoulakis, *Cosserat Continuum Mechanics*, Lecture Notes in Applied and Computational Mechanics 87, https://doi.org/10.1007/978-3-319-95156-0_1

namely 3 dofs of displacement and 3 dofs of rotation. This basic property of the Cosserat continuum has prompted Schaefer [9] to propose the use of the von Mises motor mechanics [10, 11], for the compact mathematical description of the mechanics and statics of Cosserat continua. In 1967 Schaefer remarked "Heute, im Abstand von mehr als 40 Jahren, muss man feststellen, daß von dieser Motorrechnung nur in wenigen Fällen Gebrauch gemacht worden ist; sie ist fast in Vergessenheit geraten".[2] Schaefer's complaint is still true today, more than 40 years after the publication of his paper. To this end we use here some basic concepts of motor algebra [12, 13] and motor calculus [9, 14, 15] and analysis [14], as applied to rigid body mechanics and to Cosserat continuum mechanics, with the aim to make the analogy more transparent between the micromechanics of rigid-granular media and Cosserat continuum mechanics. The analysis is restricted here to infinitesimal particle displacements and rotations. Incorporation of finite rotations [16] and introduction of non-Abelian motor calculus [17] lie outside the scope of this textbook.

The present work is meant as an addendum to a standard Continuum Mechanics course and is addressed to post graduate Students and Researchers. The reader must have been exposed to the basic concepts and notions of Continuum Mechanics [18, 19]. Some sections in the present textbook are inspired by the book of Becker and Bürger [19], that follows the German tradition of presenting the subject.

In terms of notation we use mainly Cartesian coordinates, bold face letters for vectors and the Gauss-Einstein summation convention over repeated indices. However, some sections are developed in general fixed-in-space curvilinear coordinates, in order to illustrate some fine but important details of the mathematical structure of the Cosserat continuum theory [8].

References

1. Voigt, W. (1887). Theoretische Studien über die Elastizitätsverhältnisse der Krystalle. *Abhandlungen der Mathematischen Classe der Königlichen Gesellschaft der Wissenschaften in Göttingen, 34,* 3–100.
2. Voigt, W. (1894). Über Medien ohne innere Kräfte und über eine durch sie gelieferte mechanische Deutung der Maxwell-Hertz'schen Gleichungen. *Annalen der Physik, 288,* 665–672.
3. Stojanović, R. (1970). *Recent developments in the theory of polar continua.* CISM Lectures. Springer.
4. Schaefer, H. (1967). Das Cosserat-Kontinuum. *Zeitschrift für Angewandte Mathematik und Mechanik, 47,* 485–498.
5. Hamel, G. (1921). Elementare Mechanik. *Zeitschrift für Angewandte Mathematik und Mechanik, 1*(3), 219–223.
6. Cosserat, E., & Cosserat, F. (1909). *Théorie de corps déformables.* Paris: Librairie Scientifique A. Hermann et Fils.

[2]"Today, at a distance of more than 40 years, it must be acknowledged that this motor mechanics has only been used in a few cases, it has almost fallen into oblivion".

7. Sudria, J. (1935). *L'action euclidienne de déformation et de mouvement*. Mémorial des sciences physiques. Paris: Gauthier-Villars. 56.
8. Günther, W. (1958). Zur Statik und Kinematik des Cosseratschen Kontinuums. *Abhandlungen der Braunschweigische Wissenschaftliche Gesellschaft, 10,* 195–213.
9. Schaefer, H. (1967). Analysis der Motorfelder in Cosserat-Kontinuum. *Zeitschrift für Angewandte Mathematik und Mechanik, 47,* 319–332.
10. Mises, R. V. (1924). Motorrechnung, ein neues Hilfsmittel der Mechanik. *Zeitschrift für Angewandte Mathematik und Mechanik, 4*, 155–181.
11. Mises, R. V. (1924). Anwendungen der Motorrechnung. *Zeitschrift für Angewandte Mathematik und Mechanik, 4*, 193–213.
12. Brand, L. (1940). *Vector and tensor analysis*. Wiley.
13. Talpaert, Y. R. (2003). *Mechanics, tensors and virtual works*. Cambridge International Science Publishing.
14. Kessel, S. (1967). Stress functions and loading singularities for the infinitely extended linear elastic-isotropic Cosserat Continuum. In: *Proceedings of the IUTAM-symposium on the generalized Cosserat continuum and the continuum theory of dislocations with applications*. Stuttgart: Springer.
15. Povstenko, Y. Z. (1986). Analysis of motor fields in Cosserat Continua of two and one dimensions and its applications. *Zeitschrift für Angewandte Mathematik und Mechanik, 66,* 505–507.
16. Besdo, D. (1974). Ein Beitrag zur nichtlinearen Theorie des Cosserat-Kontinuums. *Acta Mechanica, 20,* 105–131.
17. Stumpf, H., & Badur, J. (1990). On the non-Abelian motor calculus. *Zeitschrift für Angewandte Mathematik und Mechanik, 70,* 551–555.
18. Itskov, M. (1965). *Tensor algebra and tensor analysis for engineers*. Springer.
19. Becker, E., & Bürger, W. (1975). Kontinuumsmechanik. In H. Görtler (Ed.), *Leitfäden der angewandten Mathematik und Mechanik* (p. 229). Springer.

Chapter 2
Rigid-Body Mechanics and Motors

Abstract This chapter lays down the fundamental representation concepts that will be used in the book thereafter. It eventually defines the concept of a "von Mises motor", which is a compound vector including force and moment vectors. This compound representation of forces and moments in turn defines a geometric space/ representation, where all the balance laws are going to be formulated upon. It continues by laying the basic theorems that will be used to formulate the Cosserat continuum, together with the appropriate kinematic fields conjugate to the "motor" vectors that are naturally called "kinematic von Mises motors". Such a kinematic motor is a compound vector including linear velocity and spin (angular velocity), fully describing a rigid body motion in the new reduced geometric representation.

In the Mechanics literature we can find various ways of representing rigid-body statics and kinematics. Starting from elementary geometric statements taken from vector mechanics, we introduce here the concept of *motor* as applied to rigid body mechanics with the aim to reach in later chapters a better understanding of the micromechanics of Cosserat continua.

A motor is the synthesis of two words, *moment* and *vector*. The word was coined by Clifford [1] in his *Preliminary Sketch of bi-Quaternions* (1873), and was used by Richard von Mises [2, 3] in the sense given to it by Study [4] in his *Geometrie der Dynamen* (1903). As also pointed out by Schaefer, the paper of von Mises is another difficult to read reference. Note that a section on motor algebra, as an algebra of duals, and its application to rigid-body mechanics can be found in the textbook of Brand [5] and an introductory chapter related to the so-called *dynams*[1] can be found in the more recent book of Talpaert [6].

[1]In German literature this is called *Dyname,* a term that stems from the Greek word *δύναμις*; in French literature it is called *torseur.*

I. Vardoulakis, *Cosserat Continuum Mechanics*, Lecture Notes in Applied and Computational Mechanics 87, https://doi.org/10.1007/978-3-319-95156-0_2

2.1 Some Definitions from Vector Mechanics

2.1.1 Line Vectors

Geometrically, a vector is defined by the following two statements: (a) Every ordered pair of points $\{\mathrm{A}, \mathrm{B}\}$ in an Euclidian space defines a vector, denoted as $\boldsymbol{a} = \overrightarrow{\mathrm{AB}}$. (b) All ordered pairs of points that can be brought into congruence through a parallel translation define the same vector. From this definition follows that all vectors can be mapped on the set of pairs of points that result by connecting all points P in space with a common origin O. This means that in the three dimensional Euclidian space $'\mathrm{E}^3$ we have as many vectors as points, namely ∞^3.

Thus, any vector $\boldsymbol{r}$ is represented by the *point vector* $\mathbf{R} = \overrightarrow{\mathrm{OP}}$ or, for fixed Cartesian coordinates with the origin at point O, by the (3×1)-column of the coordinates of the endpoint $\mathrm{P}(x_i)$ (Fig. 2.1),

$$\boldsymbol{r} \leftrightarrow \mathbf{R} = \overrightarrow{\mathrm{OP}} = x_i \boldsymbol{e}_i \leftrightarrow \begin{Bmatrix} x_1 \\ x_2 \\ x_3 \end{Bmatrix} \tag{2.1}$$

where $\boldsymbol{e}_i$ are the Cartesian basis vectors.

Richard von Mises in his paper on *Motor Calculus*, [2, 3] has pointed to the well-known difference between a vector and a *beam*,[2] the latter being a term originally introduced by Study [4]. The definition of a beam is as follows: (a) Every ordered pair of points $\{\mathrm{A}, \mathrm{B}\}$ in space defines a beam. (b) All ordered pairs of points that can be brought into congruence through a parallel translation along their axis (ε), define the same beam.

We recognize from this definition that a typical example of a beam is the *line-force*, as this is used within the frame of rigid-body statics. Indeed, the force in rigid-body statics is not a vector in the usual sense but a *sliding vector*, i.e. a line-vector, denoted as $\vec{F}^{(\varepsilon)}$, that can slide along its axis or line of action (ε). From the definition follows that all beams can be generated, if we connect all points P in space with points O that lie on one and the same plane (E); Fig. 2.1. This means in turn that in $'\mathrm{E}^3$ we have as many beams as points in space and points on a plane, namely ∞^5.

As already mentioned above, a line force with axis (ε) is denoted as $\vec{F}^{(\varepsilon)}$. For any point $\mathrm{A} \in (\varepsilon)$ we define a *fixed force,* denoted as $\vec{F}^{\mathrm{A}}$, that is attached to the point A and corresponds to the ordered pair of points $\{\mathrm{A}, \mathrm{B}\}$; Fig. 2.2. Obviously $\vec{F}^{(\varepsilon)}$ as a sliding vector is the totality of all these point-fixed forces,

[2]German: *Stab*.

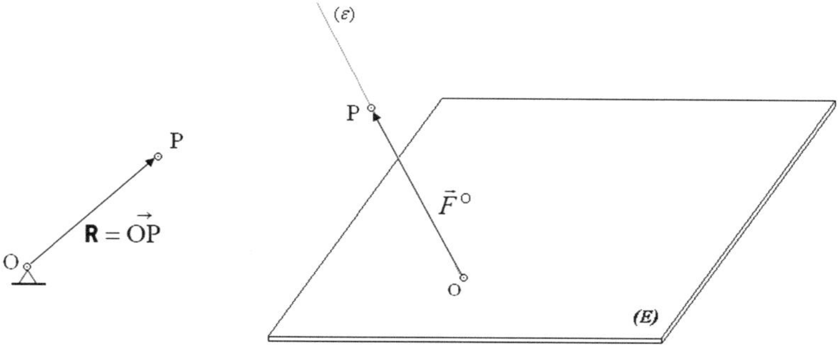

Fig. 2.1 Vector and "beam"

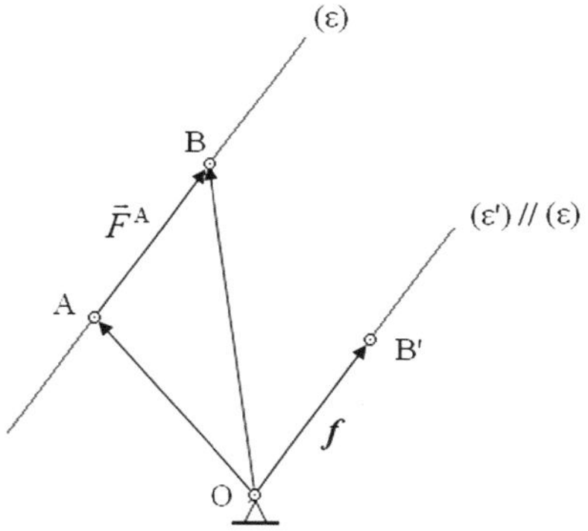

Fig. 2.2 Fixed force as a point difference

$$\vec{F}^{(\varepsilon)} = \left\{ \vec{F}^{\mathrm{A}} | \mathrm{A} \in (\varepsilon) \wedge \mathrm{B} \in (\varepsilon) : \left| \vec{F}^{\mathrm{A}} \right| = \left| \overrightarrow{\mathrm{AB}} \right| = const. \right\} \tag{2.2}$$

We can always define a vector $\boldsymbol{F}$, that can be brought into congruence with the "beam" $\vec{F}^{(\varepsilon)}$, such that $\vec{F}^{(\varepsilon)} \subset \boldsymbol{F}$. This vector $\boldsymbol{F}$ has a unique point vector representative, $\boldsymbol{f} = \overrightarrow{\mathrm{OB}'}$, that results by a parallel translation of any of the $\vec{F}^{\mathrm{A}}$ to the origin. Alternatively $\boldsymbol{f}$ can be seen as the point-vector that results as the difference between the point vectors $\overrightarrow{\mathrm{OA}}$ and $\overrightarrow{\mathrm{OB}}$ that define the endpoints of $\vec{F}^{\mathrm{A}}$,

$$\boldsymbol{f} = \overrightarrow{\mathrm{OB}'} = \overrightarrow{\mathrm{OB}} - \overrightarrow{\mathrm{OA}} = (b_i - a_i)\boldsymbol{e}_i = \boldsymbol{f}_i \boldsymbol{e}_i \tag{2.3}$$

We consider now a sliding force $\vec{F}^{(\varepsilon)}$, with axis (ε), and we select a representation of that force through the fixed force $\vec{F}^{\mathrm{A}}$, with endpoints the oriented pair $\{\mathrm{A}, \mathrm{B}\} \in (\varepsilon)$ (Fig. 2.3). The moment of the force $\vec{F}^{(\varepsilon)}$ with respect to a point O is

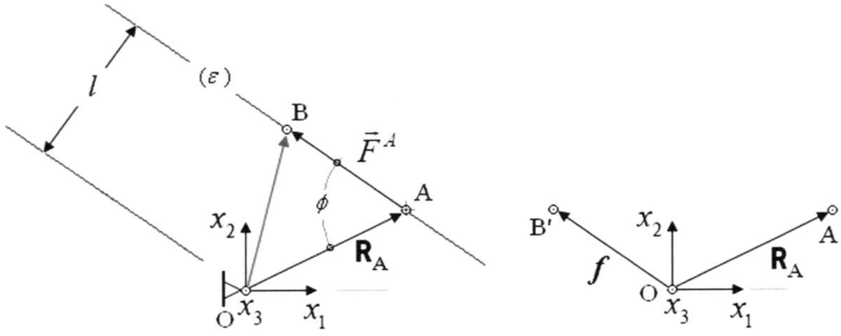

Fig. 2.3 Definition of the moment of a force

the vector product of the position vector $\overrightarrow{\mathrm{OA}}$ and the point force vector $\vec{F}^{\mathrm{A}}$, denoted as

$$\boldsymbol{M}^{\mathrm{O}} = \overrightarrow{\mathrm{OA}} \times \vec{F}^{\mathrm{A}} \tag{2.4}$$

If $\boldsymbol{f} = \overrightarrow{\mathrm{OB}'}$ is the point vector that is assigned to the fixed force $\vec{F}^{\mathrm{A}}$, Eq. (2.3), then the moment of the force $\vec{F}^{(\varepsilon)}$ with respect to point O is computed from the vector product of the position vector of the point of attachment of $\vec{F}^{\mathrm{A}}$, $\mathbf{R}_{\mathrm{A}} = \overrightarrow{OA}$, and the point vector $\boldsymbol{f} = \overrightarrow{\mathrm{OB}'}$ that is assigned to the fixed force $\vec{F}^{\mathrm{A}}$,

$$\boldsymbol{M}^{\mathrm{O}} = \overrightarrow{\mathrm{OA}} \times \vec{F}^{\mathrm{A}} = \mathbf{R}_{\mathrm{A}} \times \boldsymbol{f} = \varepsilon_{ijk}\boldsymbol{e}_i a_j f_k \tag{2.5}$$

where $\{\boldsymbol{e}_1, \boldsymbol{e}_2, \boldsymbol{e}_3\}$ is a right-handed Cartesian basis, a_i and f_i the Cartesian components of the vectors $\mathbf{R}_{\mathrm{A}} = \overrightarrow{OA}$ and $\boldsymbol{f} = \overrightarrow{\mathrm{OB}'}$, and ε_{ijk} is the corresponding Cartesian permutation tensor,

$$\varepsilon_{ijk} = \begin{cases} 1 & if: (i,j,k) = cycl(1,2,3) \\ -1 & if: (i,j,k) = cycl(2,1,3) \\ 0 & else \end{cases} \tag{2.6}$$

It can be easily seen that the moment vector $\boldsymbol{M}^{\mathrm{O}}$ depends on the choice of point O and is independent of the choice of the point A of attachment of the force, since: (a) $\boldsymbol{M}^{\mathrm{O}}$ is normal to the plane $(\mathrm{O}, \varepsilon)$, (b) $\boldsymbol{M}^{\mathrm{O}}$ is oriented in such a way that the system of vectors $\{\mathbf{R}_{\mathrm{A}}, \boldsymbol{f}, \boldsymbol{M}\}$ is right-handed. (c) The magnitude of $\boldsymbol{M}^{\mathrm{O}}$ is computed from the magnitude of the force, $\boldsymbol{F}$,

$$F = |\boldsymbol{F}| = |\vec{F}| = |\vec{F}^{(\varepsilon)}| = |\vec{F}^{\mathrm{A}}| = |\boldsymbol{f}| = \sqrt{f_i f_i} \tag{2.7}$$

The distance ℓ of the reference point O from the axis (ε) of the force is,

$$\ell = \left|\overrightarrow{\mathrm{OA}}\right| \sin \varphi \tag{2.8}$$

and,

$$\left|\boldsymbol{M}^{\mathrm{O}}\right| = \left|\overrightarrow{\mathrm{OA}}\right| |\boldsymbol{f}| \sin \phi = |\boldsymbol{f}| \ell \tag{2.9}$$

The orientation and magnitude of the moment vector follow from the geometric representation of it as a surface element vector, as shown in Fig. 2.4,

$$\left|\boldsymbol{M}^{\mathrm{O}}\right| = \overrightarrow{\mathrm{O\Gamma}} = \overrightarrow{\mathrm{OA}} \times \overrightarrow{\mathrm{OB}'} = \begin{vmatrix} \vec{e}_1 & \vec{e}_2 & \vec{e}_3 \\ a_1 & a_2 & a_2 \\ f_1 & f_2 & f_3 \end{vmatrix} \tag{2.10}$$

$$\left|\boldsymbol{M}^{\mathrm{O}}\right| = (\mathrm{OAO'B}) = (\mathrm{OA})(\mathrm{OB}) \sin \theta. \tag{2.11}$$

2.1.2 *Force-Couples*

Let $\vec{F} = \vec{F}^{(\varepsilon)}$ and $\vec{F}' = \vec{F}'^{(\varepsilon')}$ be two forces with (ε) and (ε') as their axes respectively. For any two points $\mathrm{A} \subset (\varepsilon)$ and $\mathrm{A}' \subset (\varepsilon')$, the corresponding point-fixed forces are denoted as $\vec{F}^{\mathrm{A}} \subset \vec{F}^{(\varepsilon)}$ and $\vec{F}'^{\mathrm{A}'} \subset \vec{F}'^{(\varepsilon')}$ and the corresponding force vectors are $\boldsymbol{F}$ and $\boldsymbol{F}'$. As shown in Fig. 2.5, we define a force-couple $(\vec{F}^{\mathrm{A}}, \vec{F}'^{\mathrm{A}'})$ as a set of opposite forces with parallel axes (ε) and (ε')

$$\boldsymbol{F}' = -\boldsymbol{F}, \quad (\varepsilon')//(\varepsilon) \tag{2.12}$$

By selecting an arbitrary origin O, we can compute the moment of each of the two forces that make up the considered force-couple as,

$$\boldsymbol{M}^{\mathrm{O}} = \overrightarrow{\mathrm{OA}} \times \vec{F}^{\mathrm{A}} \tag{2.13}$$

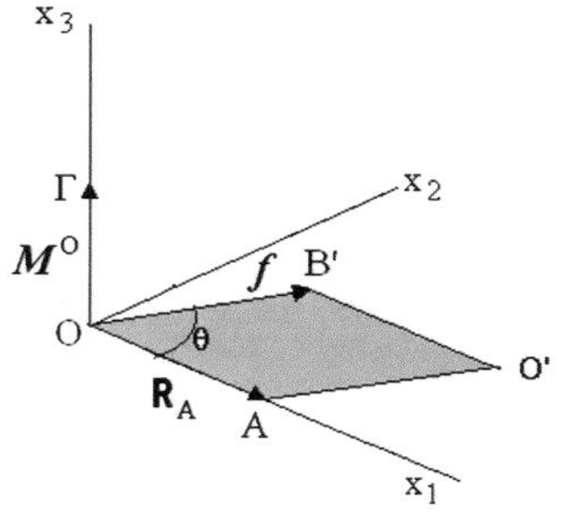

Fig. 2.4 The moment vector as a surface element vector

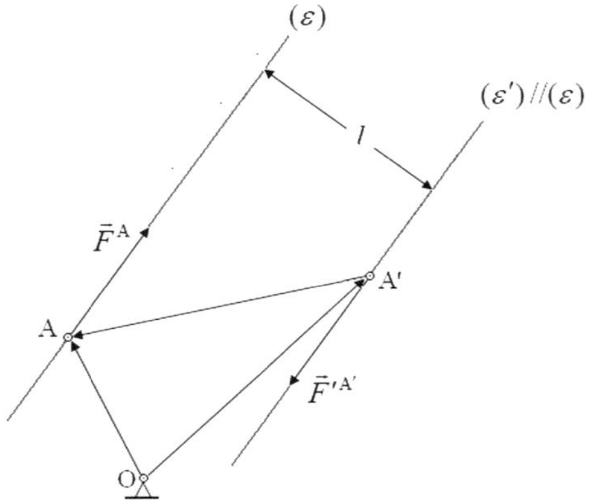

Fig. 2.5 Non-collinear force-couple

$$\boldsymbol{M}^{O'} = \overrightarrow{OA'} \times \vec{F}'^{A'} \tag{2.14}$$

We define the moment of the force-couple $(\boldsymbol{F}, -\boldsymbol{F})$ to be the sum of the moments of its components. This is a vector that lies normal to the plane $(\varepsilon, \varepsilon')$ of the force-couple and does not depend on the choice of the origin O (Fig. 2.5),

$$\boldsymbol{M} = \boldsymbol{M}^{O} + \boldsymbol{M}^{O'} = \mathbf{R}_{A} \times \boldsymbol{f} + \mathbf{R}_{A'} \times (-\boldsymbol{f}) = (\mathbf{R}_{A} - \mathbf{R}_{A'}) \times \boldsymbol{f} \tag{2.15}$$

or

$$\boldsymbol{M} = \overrightarrow{A'A} \times \vec{F}^{A} \tag{2.16}$$

with

$$M = |\boldsymbol{M}| = F\ell \tag{2.17}$$

where F is the magnitude of the force $\vec{F}$ and ℓ is the distance between the two parallel axes. From Eq. (2.17) follows that with $\ell = 0$ a co-linear couple of forces has zero moment.

We remark that two force-couples are in equilibrium, if their moments are opposite,

$$\boldsymbol{M}_1 + \boldsymbol{M}_2 = 0 \Leftrightarrow \boldsymbol{M}_1 = -\boldsymbol{M}_2 \tag{2.18}$$

On the other hand two force-couples are equivalent, if their moments are equal,

$$\boldsymbol{M}_1 = \boldsymbol{M}_2 \Leftrightarrow \boldsymbol{M}_1 - \boldsymbol{M}_2 = 0 \tag{2.19}$$

Based on these remarks, we conclude that the moments of two force-couples can be added.

2.2 Vector Statics

2.2.1 *Reduction of a System of Forces*

We start our consideration with a system of line forces $\{\vec{F}_1, \vec{F}_2, \ldots\}$ that are acting on a given rigid body B in $'\mathrm{E}^3$ along their axes (ε_i); Fig. 2.6. In order to reduce these forces into a minimal set, we select an arbitrary point O and an arbitrary plane (E) in space, such that the point O does not lie on the chosen plane (E).[3] The point O and the plane (E) are called the *reduction point* and the *reduction plane*, respectively.

We decompose the force $\vec{F}_i$ into a force $\vec{\vec{F}}{}_i^{\mathrm{O}}$, that is attached to point O, and a component $\vec{F}'_i$ that lies on a line $(\varepsilon'_i) \subset (\mathrm{E})$, that results as the intersection of the plane (E) and the plane $(\Pi) \equiv (\mathrm{O}, \varepsilon_i)$; Fig. 2.7. Line (ε'_i) intersects line (ε_i) at point O′, and with that $\vec{F}_i'^{\mathrm{O}'} \subset \vec{F}'_i$. The axis of $\vec{\vec{F}}{}_i^{\mathrm{O}}$ is the line $(\bar{\varepsilon}_i) = (\mathrm{OO}')$, thus $\vec{\vec{F}}{}_i^{\mathrm{O}} \subset \vec{\vec{F}}_i$. In that sense, the decomposition of force $\vec{F}_i$ is unique,

$$\vec{F}_i = \vec{\vec{F}}_i + \vec{F}'_i \tag{2.20}$$

The above procedure can be repeated for all forces of the considered system $\{\vec{F}_1, \vec{F}_2, \ldots\}$.

Let $\vec{\vec{F}}{}^{\mathrm{O}}$ be the resultant of all components $\vec{\vec{F}}{}_i^{\mathrm{O}}$, that are set to be attached to point O (Fig. 2.8),

$$\vec{\vec{F}}{}^{\mathrm{O}} = \sum_i \vec{\vec{F}}{}_i^{\mathrm{O}} \tag{2.21}$$

The resultant of all components $\vec{F}'_i$ with axis in the plane (E) may be either a single force $\vec{F}'$ or a force-couple $(\vec{F}'^{\mathrm{A}}, -\vec{F}''^{\mathrm{B}})$, such that, $\boldsymbol{F}'' = -\boldsymbol{F}'$. As already said, the moment $\vec{\boldsymbol{M}}'$ of a force-couple does not depend upon the choice of the origin and is said to be a *"free" vector*, meaning that $\boldsymbol{M}'$ is a vector, since its axis is restricted only to be perpendicular to the plane (E),

$$\boldsymbol{M}' = \overrightarrow{\mathrm{BA}} \times \vec{F}'^{\mathrm{A}} \tag{2.22}$$

Thus, as a result of the arbitrary choice of the reduction point O and of the reduction plane (E), the system of forces $\{\vec{F}_1, \vec{F}_2, \ldots\}$ is reduced into ether: (a) A system of two skew forces, $\{\vec{\vec{F}}{}^{\mathrm{O}}, \vec{F}'\}$, with the axis of $\vec{\vec{F}}{}^{\mathrm{O}}$ passing through point O

---

[3]This section is mostly inspired by the presentation of the subject, given by my Teacher, late Professor Bitsakos (Γεωργικόπουλος Κ.Χ. και Μπιτσάκος Λ.Ι., Τεχνική Μηχανική Β', Γραφοστατική, Εκδ. Τεχνικού Επιμελητηρίου της Ελλάδος, 1967.).

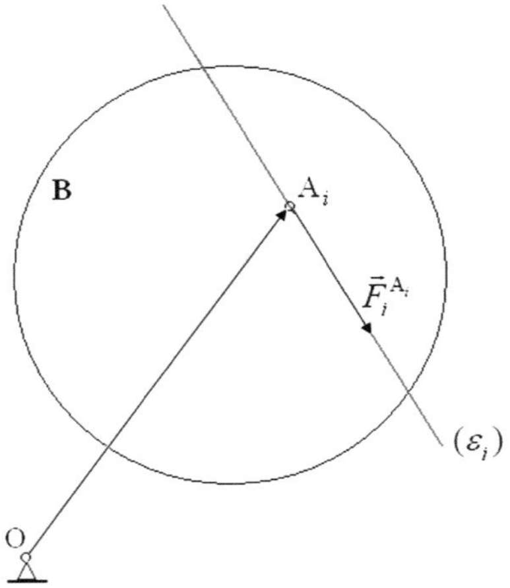

Fig. 2.6 System of line forces acting on a rigid body

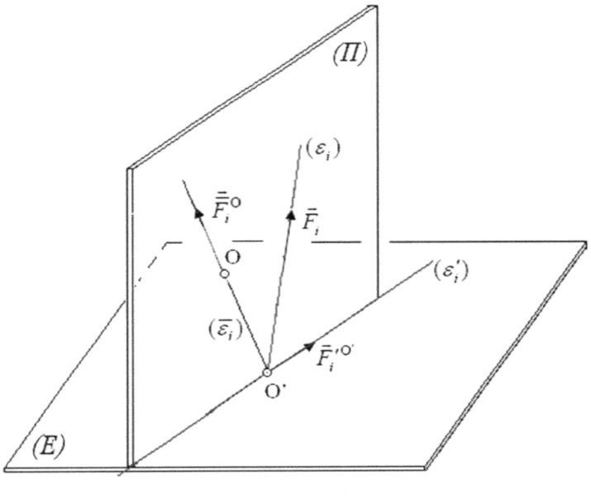

Fig. 2.7 Decomposition of a force into a force passing through a point O and a force lying in a plane (E)

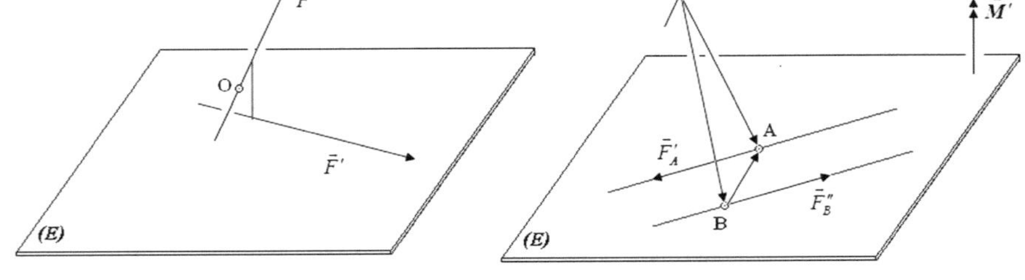

Fig. 2.8 Reduction of a force system into two a force passing through the reduction point O and **a** a skew force or **b** a force-couple, lying in the reduction plane (E)

and the axis of $\vec{F}'$ lying in the plane (E), or (b) a system of a force $\vec{F}^{\mathrm{O}}$ and a couple $\boldsymbol{M}'$, denoted as $\{\vec{F}^{\mathrm{O}}, \boldsymbol{M}'\}$, with the axis of $\vec{F}^{\mathrm{O}}$ passing through point O and the couple $\vec{\boldsymbol{M}}'$ being normal to the plane (E).

For the given system of forces $\{\vec{F}_1, \vec{F}_2, \ldots\}$, the above described reduction strategy introduces an equivalence class of ∞^6 reduced systems $\{\vec{F}^{\mathrm{O}}, \vec{F}'\}$ or $\{\vec{F}^{\mathrm{O}}, \boldsymbol{M}'\}$; their multitude being determined by the multitude of points O and planes (E) in $'\mathrm{E}^3$.

The system of the two skew forces $\{\vec{F}^{\mathrm{O}}, \vec{F}'\}$ can be further reduced into a force passing through point O and a force-couple as follows (Fig. 2.9): Let the axis of the force $\vec{F}'$ in the plane (E) be line (ε'). From point O we draw a line (ε'') that is parallel to (ε') and along this line we add at point O the self-equilibrating force-doublet $(\vec{F}''^{\mathrm{O}}, -\vec{F}''^{\mathrm{O}})$ such that the vectors that correspond to the forces $\vec{F}'$ and $\vec{F}''^{\mathrm{O}}$ are equal (Fig. 2.9) to a vector F'. With this construction the original pair of skew forces $\{\vec{F}^{\mathrm{O}}, \vec{F}^{(\mathrm{E})}\}$ is replaced by the resultant force,

$$\vec{S}^{\mathrm{O}} = \vec{F}^{\mathrm{O}} + \vec{F}''^{\mathrm{O}} \tag{2.23}$$

that lies in the plane $(\varepsilon, \varepsilon'')$ and is passing through point O, and the force-couple $(\vec{F}', -\vec{F}'')$ that is made of the force $\vec{F}'$ with axis line (ε') in the plane (E) and the force $-\vec{F}''$, with axis line $(\varepsilon'')//(\varepsilon')$, passing through point O. The moment $\boldsymbol{M}''$ of this force-couple is a vector, that lies normal to the plane $(\mathrm{E}') = (\mathrm{O}, \varepsilon')$,

$$\boldsymbol{M}'' = \overrightarrow{\mathrm{OO}'} \times \vec{\boldsymbol{F}}' \tag{2.24}$$

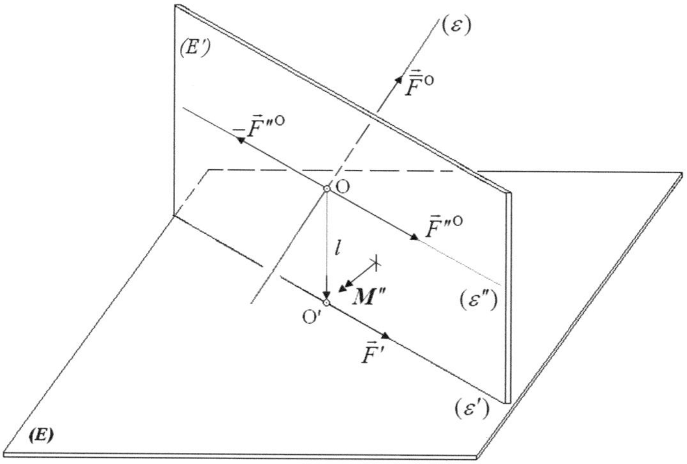

Fig. 2.9 Further reduction of a system of two skew forces

Note that in both cases the force that is attached to the reduction point O is equal to the resultant force of the considered system,

$$\vec{S}^{\mathrm{O}} = \vec{F}^{\mathrm{O}} + \vec{F}''^{\mathrm{O}} \tag{2.25}$$

or

$$\vec{S}^{\mathrm{O}} = \vec{F}^{\mathrm{O}} \tag{2.26}$$

The above considerations resulted in the following general theorem:

Given is a system of forces $\{\vec{F}_1, \vec{F}_2, \ldots\}$ that are acting on a rigid body, a plane (E) and a point O outside this plane. The given system of forces can always be reduced into a force that is passing through the point O and is equal to the resultant force of the system, $\vec{F}^{\mathrm{O}}$, and a force-couple $\boldsymbol{M}^{(\mathrm{E})}$ that is either normal to the plane (E) ($\boldsymbol{M}^{(\mathrm{E})} \equiv \boldsymbol{M}'$) or is normal to the plane (E$'$) that is made of point O and the axis of the resultant force in the plane (E) ($\boldsymbol{M}^{(\mathrm{E})} \equiv \boldsymbol{M}''$).

We note that if the resultant force and the resultant force-couple are coplanar, then the vector product of the corresponding vectors vanishes,

$$\boldsymbol{S} \cdot \boldsymbol{M} = \boldsymbol{0} \tag{2.27}$$

In this case the original system of forces can be reduced into a single force as indicated in Fig. 2.10.

In general a system of forces $\{\vec{F}_1, \vec{F}_2, \ldots\}$ is reduced into a single force, if there exists a point O in space, such that the moment of these forces with respect to this point vanishes (Fig. 2.11),

$$\boldsymbol{M}^{\mathrm{O}} = \sum_i \overrightarrow{\mathrm{OA}_i} \times \vec{\boldsymbol{F}}_i^{\mathrm{A}_i} = \boldsymbol{0} \tag{2.28}$$

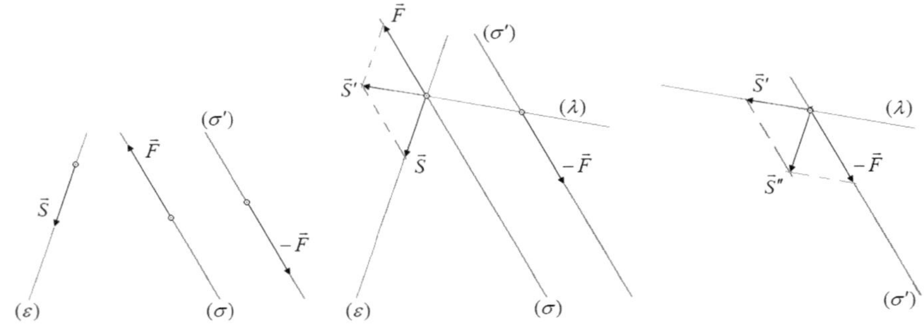

Fig. 2.10 Synthesis of a coplanar system, consisting of a force and force-couple, into a single resultant force

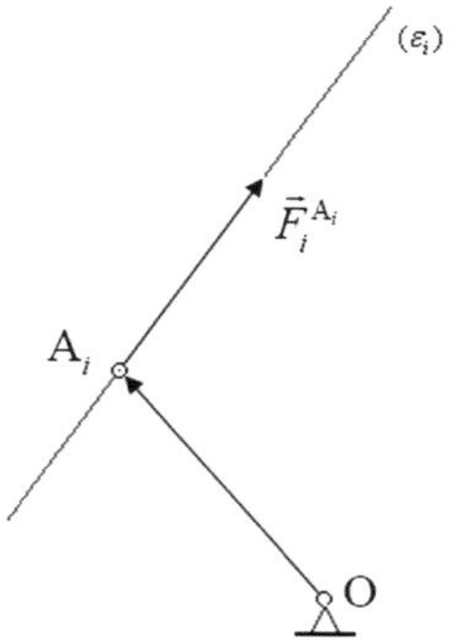

Fig. 2.11 Geometric layout for the formulation of the equilibrium equations

In this case the system of forces is said to be in equilibrium, if in addition the resultant force of that system vanishes,

$$\boldsymbol{S} = \sum_i \boldsymbol{F}_i = \boldsymbol{0} \tag{2.29}$$

2.2.2 *Transport Law of Forces: The Dynamic Motor*

Let us consider a fixed force $\vec{F}^{\mathrm{P}}$ acting at a point P of its axis (ε); Fig. 2.12. We denote the corresponding point vector by $\boldsymbol{F}^{\mathrm{P}}$. If we consider now a force that arises through parallel translation of $\vec{F}^{\mathrm{P}}$ to another point O_1, resulting to the force $\vec{F}^{\mathrm{O}_1}$ with axis $(\varepsilon_1)//(\varepsilon)$, then

$$\boldsymbol{F}^{\mathrm{O}_1} = \boldsymbol{F}^{\mathrm{P}} \tag{2.30}$$

The force $\vec{F}^{\mathrm{P}}$ can be replaced by the force $\vec{F}^{\mathrm{O}_1}$ and a co-planar force-couple $(\vec{F}^{\mathrm{P}}, -\vec{F}^{\mathrm{O}_1})$ with moment,

$$\boldsymbol{M}^{\mathrm{O}_1} = \overrightarrow{\mathrm{O}_1\mathrm{P}} \times \vec{F}^{\mathrm{P}} \tag{2.31}$$

We note that for all points $\mathrm{P}' \subset (\varepsilon)$ along the axis of $\vec{F}^{\mathrm{P}}$,

$$\boldsymbol{M}^{\mathrm{P}'} = \overrightarrow{\mathrm{P}'\mathrm{P}} \times \vec{F}^{\mathrm{P}} = \boldsymbol{0} \tag{2.32}$$

Thus, by construction the systems $\{\vec{F}^{\mathrm{O}_1}, \boldsymbol{M}^{\mathrm{O}_1}\}$ are identical for all $\mathrm{O}_1 \subset (\varepsilon_1)$ and are all reducible to the original system $\{\vec{F}^{\mathrm{P}}, \boldsymbol{M}^{\mathrm{P}'} = \boldsymbol{0}\}$ for all $\mathrm{P}' \subset (\varepsilon)$. In this case we say that the given force is transported from point P to point O_1.

We may now select another reference point O_2 with,

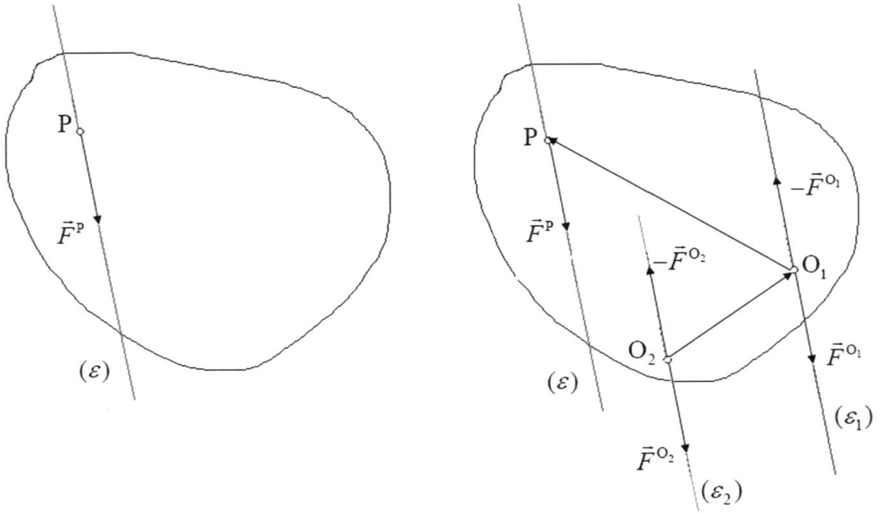

Fig. 2.12 The transport law of a line force

$$\boldsymbol{M}^{\mathrm{O}_2} = \overrightarrow{\mathrm{O}_2\mathrm{P}} \times \vec{F}^{\mathrm{P}} = \left(\overrightarrow{\mathrm{O}_2\mathrm{O}_1} + \overrightarrow{\mathrm{O}_1\mathrm{P}}\right) \times \vec{F}^{\mathrm{P}} = \boldsymbol{M}^{\mathrm{O}_1} + \overrightarrow{\mathrm{O}_2\mathrm{O}_1} \times \vec{F}^{\mathrm{O}_1} \tag{2.33}$$

The above results are summarized into the following *transport law* for a single line force, Eqs. (2.30) and (2.33),

$$\boldsymbol{F}^{\mathrm{O}_2} = \boldsymbol{F}^{\mathrm{O}_2} = \boldsymbol{F} \tag{2.34}$$

$$\boldsymbol{M}^{\mathrm{O}_2} = \boldsymbol{M}^{\mathrm{O}_1} + \overrightarrow{\mathrm{O}_2\mathrm{O}_1} \times \vec{F}^{\mathrm{O}_1} \tag{2.35}$$

Equation (2.34) means equality of the force vectors that correspond to the point vectors acting at arbitrary points $\mathrm{O}_1 \subset (\varepsilon_1)$ and $\mathrm{O}_2 \subset (\varepsilon_2)$ along the axes $(\varepsilon_1)//(\varepsilon_2)$. We remark that in the considered case, of the transport of a single line force, we have always that,

$$\boldsymbol{F} \cdot \boldsymbol{M}^{\mathrm{O}} = \boldsymbol{0} \tag{2.36}$$

The vector-moment compound,

$$\underline{p} = \begin{pmatrix} \boldsymbol{F} \\ \boldsymbol{M}^{\mathrm{O}} \end{pmatrix} \tag{2.37}$$

is called a *proper von Mises motor,* if the force $\boldsymbol{F}$ and the couple $\boldsymbol{M}^{\mathrm{O}}$ obey the transport law, Eqs. (2.34) and (2.35), and the normality condition, Eq. (2.36).

Any system of forces acting on a rigid body can be reduced to a single resultant force $\vec{S}^{\mathrm{O}}$ and a couple $\boldsymbol{M}^{(\mathrm{E})}$ by the choice of a plane (E) and a point O outside that plane. If we select a different reduction point, say point O_1, then the given system is reduced to the force $\vec{S}^{\mathrm{O}_1}$ that obeys the transport law, Eq. (2.30),

$$\boldsymbol{S}^{\mathrm{O}_1} = \boldsymbol{S}^{\mathrm{O}_2} = \boldsymbol{S} \tag{2.38}$$

and to the couple,[4]

$$\boldsymbol{M}^{\mathrm{O}_1} = \boldsymbol{M}^{(\mathrm{E})} + \overrightarrow{\mathrm{O}_1\mathrm{O}} \times \vec{S}^{\mathrm{O}} \tag{2.39}$$

We observe that in this case the normality condition, Eq. (2.36), does not apply necessarily.

A compound of the two vectors,

$$\underline{p} = \begin{pmatrix} \boldsymbol{S} \\ \boldsymbol{M}^{\mathrm{O}} \end{pmatrix} \tag{2.40}$$

will be called a *von Mises motor*, if the vectors $\boldsymbol{S}$ and $\boldsymbol{M}^{\mathrm{O}}$ fulfil the transport law, Eqs. (2.34) and (2.35). In particular the motor, defined above, is called a *dynamic motor*.[5] We have shown that a system of forces acting on a rigid body is always reducible into a dynamic motor.

2.2.3 Central Axis of a System of Forces and Axis of a Motor

Consider a system of forces $\{\vec{F}_1, \vec{F}_2, \ldots\}$ acting on a rigid body and let $\{\boldsymbol{F}_1; \boldsymbol{F}_2, \ldots\}$ be the corresponding system of (free) force vectors. Let the resultant force vector be,

$$\boldsymbol{S} = \sum_i \boldsymbol{F}_i \tag{2.41}$$

Following the above described procedure, the system may be reduced to a fixed force $\vec{S}^{\mathrm{O}}$ attached to a point O and a couple $\boldsymbol{M}^{(\mathrm{E})}$. Consider two points O_1 and O_2. According to Eq. (2.39), if we transport the force $\boldsymbol{S}$ to these points, then the moments of the system of forces in reference to these points are,

[4]This is true because two force-couples can be added by adding their moments.

[5]Δύναμις, Greek for dynamic action.

$$\boldsymbol{M}^{\mathrm{O}_1} = \boldsymbol{M}^{(\mathrm{E})} + \overrightarrow{\mathrm{O}_1\mathrm{O}} \times \vec{S}^{\mathrm{O}} \tag{2.42}$$

$$\boldsymbol{M}^{\mathrm{O}_2} = \boldsymbol{M}^{(\mathrm{E})} + \overrightarrow{\mathrm{O}_2\mathrm{O}} \times \vec{S}^{\mathrm{O}} \tag{2.43}$$

Their difference is independent of $\boldsymbol{M}^{(\mathrm{E})}$,

$$\boldsymbol{M}^{\mathrm{O}_2} - \boldsymbol{M}^{\mathrm{O}_1} = \left(\overrightarrow{\mathrm{O}_2\mathrm{O}} - \overrightarrow{\mathrm{O}_1\mathrm{O}}\right) \times \vec{S}^{\mathrm{O}} = \overrightarrow{\mathrm{O}_2\mathrm{O}_1} \times \boldsymbol{S} \tag{2.44}$$

Thus

$$\boldsymbol{M}^{\mathrm{O}_2} = \boldsymbol{M}^{\mathrm{O}_1} + \mathbf{R}_{21} \times \boldsymbol{S} \tag{2.45}$$

We remark that the condition,

$$\boldsymbol{M}^{\mathrm{O}_2} = \boldsymbol{M}^{\mathrm{O}_1} \tag{2.46}$$

implies either of the following two possibilities: (a) $\boldsymbol{S} = \boldsymbol{0}$; i.e. the original system is equivalent to a planar force-couple. (b) The line $(\mathrm{O}_1\mathrm{O}_2)$ is parallel to the axis of resultant force. With

$$(\mathbf{R}_{21} \times \boldsymbol{S}) \cdot \boldsymbol{S} = \mathbf{R}_{21}(\boldsymbol{S} \times \boldsymbol{S}) = 0 \tag{2.47}$$

from Eq. (2.45) we get,

$$\boldsymbol{M}^{\mathrm{O}_2} \cdot \boldsymbol{S} = \boldsymbol{M}^{\mathrm{O}_1} \cdot \boldsymbol{S} \tag{2.48}$$

Thus for a given system of forces $\{\vec{F}_1, \vec{F}_2, \ldots\}$, besides their resultant force $\boldsymbol{S} = \sum_i \boldsymbol{F}_i$, invariant with respect to changes in the position of the origin is also the projection of the resultant moment vector on $\boldsymbol{S}$.

Let $\boldsymbol{e}$ and $\boldsymbol{s}$ be unit vectors with the same direction as the resultant force vector $\boldsymbol{S}$ and the point-difference vector $\mathbf{R}_{21} = \overrightarrow{\mathrm{O}_2\mathrm{O}_1}$, respectively

$$\boldsymbol{s} = \frac{\boldsymbol{S}}{\sqrt{\boldsymbol{S} \cdot \boldsymbol{S}}}, \ |s| = 1; \quad \sqrt{\boldsymbol{S} \cdot \boldsymbol{S}} = S > 0 \tag{2.49}$$

$$\boldsymbol{r} = \frac{\mathbf{R}_{12}}{\sqrt{\mathbf{R}_{12} \cdot \mathbf{R}_{12}}}, |\boldsymbol{r}| = 1; \quad \sqrt{\mathbf{R}_{12} \cdot \mathbf{R}_{12}} = r_{12} > 0 \tag{2.50}$$

From Eq. (2.45) we get

$$\boldsymbol{s} \times \boldsymbol{M}^{\mathrm{O}_2} = \boldsymbol{s} \times \boldsymbol{M}^{\mathrm{O}_1} + \boldsymbol{s} \times (\mathbf{R}_{21} \times \boldsymbol{S}) = \boldsymbol{s} \times \boldsymbol{M}^{\mathrm{O}_1} - \boldsymbol{s} \times (\boldsymbol{r} \times \boldsymbol{s}) r_{12} S \tag{2.51}$$

We assume now that the points O_1 and O_2 are on a plane (Π) that is normal to $\boldsymbol{S}$ (Fig. 2.13). At point O_1 we attach a right-handed ortho-normal basis $(\boldsymbol{s}, \boldsymbol{r}, \boldsymbol{n})$. With

$$\boldsymbol{s} \times (\boldsymbol{r} \times \boldsymbol{s}) = \boldsymbol{s} \times (-\boldsymbol{n}) = -(-\boldsymbol{r}) = \boldsymbol{r} \tag{2.52}$$

we get from Eq. (2.51)

$$\boldsymbol{s} \times \boldsymbol{M}^{O_2} = \boldsymbol{s} \times \boldsymbol{M}^{O_1} - \boldsymbol{r}\, r_{12} S = \boldsymbol{s} \times \boldsymbol{M}^{O_1} - \mathbf{R}_{12} S \tag{2.53}$$

or

$$\boldsymbol{S} \times \boldsymbol{M}^{O_2} = \boldsymbol{S} \times \boldsymbol{M}^{O_1} - \mathbf{R}_{12} S^2 \tag{2.54}$$

Consider the equation,

$$\mathbf{R}_{12} = \frac{1}{S^2} \boldsymbol{S} \times \boldsymbol{M}^{O_1} \tag{2.55}$$

This allows to determine a point $O_2 \subset (\Pi)$ such that,

$$\boldsymbol{S} \times \boldsymbol{M}^{O_2} = \boldsymbol{0} \tag{2.56}$$

This means that the moment of the considered system of forces with respect to that point O_2, $\boldsymbol{M}^{O_2} = \sum_i \overrightarrow{O_2 A_i} \times \boldsymbol{F}_i^{A_i}$, is a vector that is parallel to the resultant $\boldsymbol{S} = \sum_i \boldsymbol{F}_i$, i.e. also normal to the plane (Π). The same property holds for all points

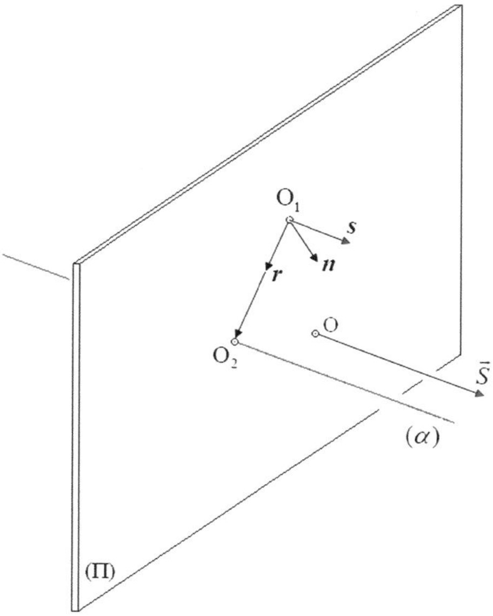

Fig. 2.13 Construction of the central axis

along an axis (α) that is normal to the plane (Π) and passes through the point O_2. For all these points $P \subset (\alpha)$ the original system is reduced into a force resultant $\boldsymbol{S}$ and a planar force-couple with its plane normal to $\boldsymbol{S}$. The corresponding motor has the property,

$$\underline{p} = \begin{pmatrix} \boldsymbol{S} \\ \boldsymbol{M}^{\mathrm{P}} \end{pmatrix} \quad \mathrm{P} \subset (\alpha) : \boldsymbol{S} \times \boldsymbol{M}^{\mathrm{P}} = 0 \tag{2.57}$$

Such an axis (α) is called the *central axis* of the given system of forces or simply the axis of the corresponding motor. From Eq. (2.55) we get the vector equation for the axis (α) of the motor, Eq. (2.57):

$$\overrightarrow{\mathrm{O_1 O_2}} \times \boldsymbol{S} = \frac{1}{\boldsymbol{S}^2} \left(\boldsymbol{S} \times \boldsymbol{M}^{\mathrm{O_1}} \right) \times \boldsymbol{S} \tag{2.58}$$

2.3 Vector Kinematics

2.3.1 Rigid-Body Motion

We summarize here some of the basic definitions and theorems of finite rigid-body kinematics that are proven in standard books of classical mechanics.

Definition A point of a body is called a *fixed point,* if after the application of the motion this point is mapped onto itself.

Theorem 1 *If a motion has four fixed points, that are not on the same plane, then the motion is an identity mapping of all points onto themselves.*

Theorem 2 *If a motion has a fixed plane and the motion is not the identity mapping, then this motion is not a real motion; it is a pseudo-motion that corresponds to a reflection of all points of the considered body with respect to the given fixed plane.*

In other words if we exclude pseudo-motions we have,

Theorem 3 *The position of all points of a rigid body is determined by the position of three of its points, provided the points are not collinear.*

Theorem 4 *If a motion possesses a fixed straight line, then this motion is a rotation with respect to that line.*

Theorem of Euler: Rotation about a single fixed point is equivalent to a rotation about an axis that is passing through this point.

If no constraints are attached to the body, then it is said to be *free.* Free rigid body kinematics are summarized in the famous,

Theorem of Chasles [7]: A rigid body can be displaced from one arbitrary position to another by means of one translation and one rotation about an axis.

In general this can be done in infinitely many ways, but the axes of rotation will always be parallel and the angles of rotation equal, if the axes have the same sense.

By some authors [8] this theorem is originally attributed to Mozzi [9] and is considered as the basis of the mechanical theory of screws [10].

If a body makes a translation and then a rotation about an axis parallel to the translation, then the body is said to have made a *twist*.

Theorem 5 *A body can always be displaced from one arbitrary position to another by means of a twist and this can be done in only one way.*

2.3.2 *Instantaneous Rigid Body Motion: The Kinematic Motor*

Let

$$\vec{\mathrm{OP}} = \mathbf{R}(X, t) \tag{2.59}$$

be the position vector of a material point $X \in \mathbf{B}$ as function of time, measured with respect to a fixed-in-space origin O. In time $t' = t + \Delta t$ the material point is moved to a new position,

$$\vec{\mathrm{OP'}} = \mathbf{R}(X, t') = \mathbf{R}(X, t + \Delta t) \approx \mathbf{R}(X, t) + \frac{\partial \mathbf{R}}{\partial t}\Delta t \tag{2.60}$$

Since the body is rigid, the distance between two arbitrary points P and Q remains constant. In general the symbol X will be omitted from the argument list of the position vector, meaning that, if not otherwise explicitly stated, the vector $\mathbf{R}(t)$ will always follow the same material point. In that sense the velocity of the material point X, which at time t was at point P, is given by

$$\boldsymbol{v} = \lim_{\Delta t \to 0} \frac{\vec{\mathrm{OP'}} - \vec{\mathrm{OP}}}{\Delta t} = \frac{d\mathbf{R}}{dt} \tag{2.61}$$

A rigid-body motion is called a *translation*, when the velocity is the same for all points of the considered body. In this case all points of the body are displaced equally during a given time interval (Fig. 2.14),

$$\vec{\mathrm{PP'}} = d\mathbf{R} = \boldsymbol{v}dt \quad \boldsymbol{v} = const. \forall X \in \boldsymbol{B} \tag{2.62}$$

In general however, the velocity will be not the same for all points of a moving rigid body. For example, if the rigid body is rotated around a fixed-in-space axis (α), then all its material points will move along circles, with their centers on that axis (Fig. 2.15). At the instantaneous position P of a material point at time t we introduce the corresponding polar basis vectors, $\boldsymbol{e}_r$ and $\boldsymbol{e}_\varphi$. The velocity vector is then purely circumferential,

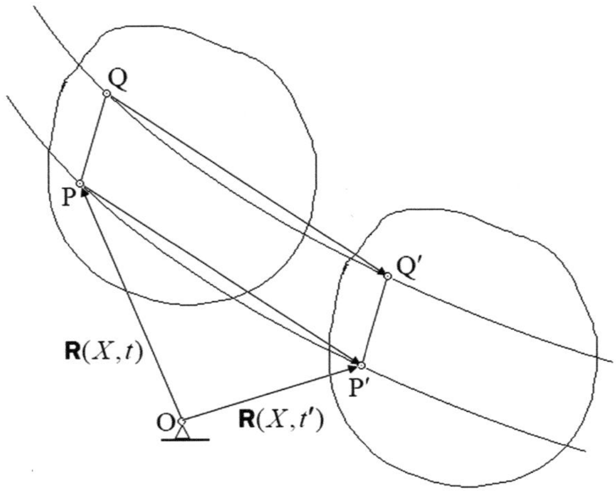

Fig. 2.14 Translatory motion of a rigid body

Fig. 2.15 Rotation around a fixed axis

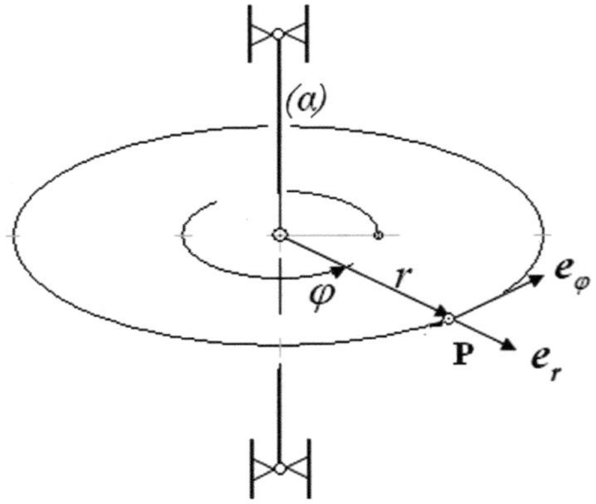

$$\boldsymbol{v}_{\mathrm{P}} = v_{\varphi}\boldsymbol{e}_{\varphi} \tag{2.63}$$

where

$$v_{\varphi} = r\omega, \quad \omega = \frac{d\varphi}{dt} \tag{2.64}$$

r is the radial distance of the point P from the axis and ω is its angular velocity or spin.

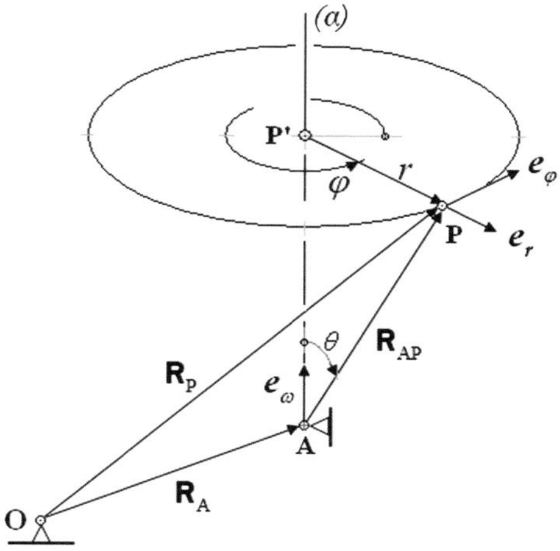

Fig. 2.16 Rotation around a fixed point

In view of Euler's Theorem, in case the motion is taking place around one fixed point A, we consider that at any instant this motion will be a rotation around an instantaneous axis (α), that will be passing through the fixed point A. If $\boldsymbol{e}_\omega$ is the unit vector along this axis of rotation, pointing in the positive sense, then according to Fig. 2.16 we have,

$$r = (\mathrm{P'P}) = |\mathbf{R}_{\mathrm{AP}}| \sin\theta = |\boldsymbol{e}_\omega \times \mathbf{R}_{\mathrm{AP}}| \tag{2.65}$$

and

$$d\mathbf{R}_{\mathrm{P}} = r d\varphi \boldsymbol{e}_\varphi \tag{2.66}$$

Thus,

$$d\mathbf{R}_{\mathrm{P}} = (\boldsymbol{e}_\omega \times \mathbf{R}_{\mathrm{AP}}) d\varphi \tag{2.67}$$

One can introduce an infinitesimal rotation vector $d\boldsymbol{\varphi}$,

$$d\boldsymbol{\varphi} = d\varphi \boldsymbol{e}_\omega \tag{2.68}$$

and the spin vector $\boldsymbol{w}$,

$$\boldsymbol{w} = \frac{d\boldsymbol{\varphi}}{dt} = \omega \boldsymbol{e}_\omega \tag{2.69}$$

In that case, we get from Eq. (2.66),

$$\boldsymbol{v}_{\mathrm{P}} = \frac{d\mathbf{R}_{\mathrm{P}}}{dt} = \boldsymbol{w} \times \mathbf{R}_{\mathrm{AP}} \tag{2.70}$$

At this point it should be emphasized that the infinitesimal rotation $d\boldsymbol{\varphi}$ and the corresponding spin $\boldsymbol{w} = d\boldsymbol{\varphi}/dt$ are vectors. This is not true however for finite rotations.

In the most general case of a free rigid-body motion, we select an arbitrary point M of the rigid body and we place a coordinate system $A(x_i')$ fixed on that point and with its axes always parallel to the axes of the fixed-in-space coordinate system. Since the point M is fixed relative to the system $A(x_i')$, the motion of the body relative to that system is a rotation about A. The instantaneous relative motion will be therefore an instantaneous rotation about an axis (α) through A.

Let P be an arbitrary point of the considered body. The velocity $\boldsymbol{v}^{\mathrm{P}}$ of this point, with respect to the fixed-in-space system $O(x_i)$, can be seen as the sum of the velocity $\boldsymbol{v}^{\mathrm{A}}$ of point A, with respect to the fixed-in-space system $O(x_i)$, and the relative velocity $\boldsymbol{v}^{\mathrm{AP}}$ of point P, with respect to point A,

$$\boldsymbol{v}^{\mathrm{P}} = \boldsymbol{v}^{\mathrm{A}} + \boldsymbol{v}^{\mathrm{AP}} \tag{2.71}$$

Equation (2.71) is illustrated in Fig. 2.17, where the velocity decomposition is shown for a plane rigid-body motion.

According to Eq. (2.70), the relative motion is a rotation around an axis $(\alpha) \supset \mathrm{A}$ with spin vector $\boldsymbol{w}$,

$$\boldsymbol{v}^{AP} = \boldsymbol{w} \times \mathbf{R}_{\mathrm{AP}} \tag{2.72}$$

Thus,

$$\boldsymbol{v}^{\mathrm{P}} = \boldsymbol{v}^{\mathrm{A}} + \boldsymbol{w} \times \mathbf{R}_{\mathrm{AP}} = \boldsymbol{v}^{\mathrm{A}} - \mathbf{R}_{\mathrm{AP}} \times \boldsymbol{w} = \boldsymbol{v}^{\mathrm{A}} + \mathbf{R}_{\mathrm{PA}} \times \boldsymbol{w} \tag{2.73}$$

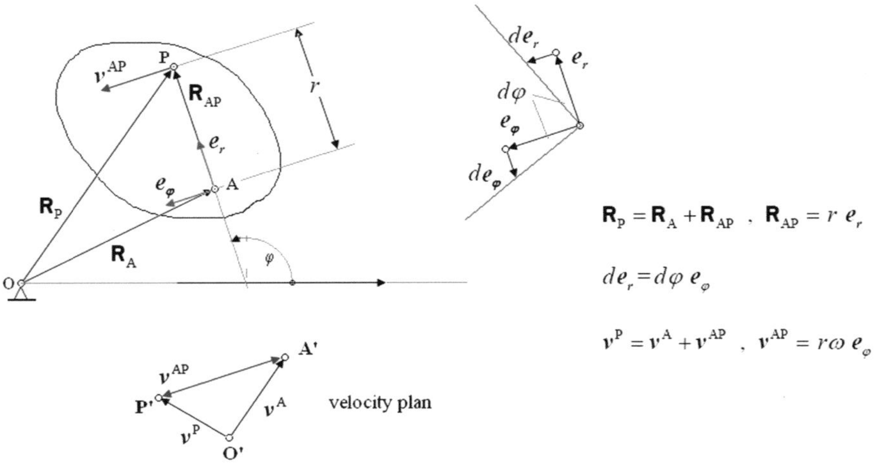

Fig. 2.17 Plane rigid-body motion and velocity plan

With,

$$\mathbf{R}_{PA} = \mathbf{R}_{A} - \mathbf{R}_{P} = \overrightarrow{PA} \tag{2.74}$$

we get,

$$\boldsymbol{v}^{P} = \boldsymbol{v}^{A} + \overrightarrow{PA} \times \boldsymbol{w} \tag{2.75}$$

According to Chasles' theorem, if we select another reference point, say B, the instantaneous axis of rotation (β) will be parallel to the line (α) and for any infinitesimal transition the angle of rotation $d\varphi$ will be also the same. This means in turn that the spin vector is invariant to the selection of the reference point,

$$\boldsymbol{w}^{B} = \boldsymbol{w}^{A} = \boldsymbol{w} \tag{2.76}$$

With respect to reference point B, the velocity of the arbitrary point P is computed in analogy to Eq. (2.75),

$$\boldsymbol{v}^{P} = \boldsymbol{v}^{B} + \overrightarrow{PB} \times \boldsymbol{w} \tag{2.77}$$

If we subtract Eqs. (2.77) and (2.75), and utilize Eq. (2.76), we get the well-known transport-law for the velocity in rigid-body kinematics,

$$\boldsymbol{w}^{B} = \boldsymbol{w}^{A} = \boldsymbol{w} \tag{2.78}$$

$$\boldsymbol{v}^{B} = \boldsymbol{v}^{A} + \overrightarrow{BA} \times \boldsymbol{w} \tag{2.79}$$

The compound of the two vectors,

$$\bar{\kappa} = \begin{pmatrix} \boldsymbol{w} \\ \boldsymbol{v}^{A} \end{pmatrix} \tag{2.80}$$

will be called a *kinematic von Mises motor*, if the vectors $\boldsymbol{w}^{A}$ and v^{A} fulfil the transport law, Eqs. (2.78) and (2.79). We have shown that the velocities of the particles of a rigid body constitute a kinematic motor. We remark that in case of a plane motion (Fig. 2.17) we have that,

$$\boldsymbol{w} \cdot \boldsymbol{v}^{A} = \boldsymbol{0} \tag{2.81}$$

In this case $\bar{\kappa}$ is a *proper von Mises motor*. However, as we see in the next section, this not the only possibility.

2.3.3 Central Axis of Rotation: Twist

The above reduction allows us to transfer the result concerning the central axis of a system of forces directly to the characterization of rigid-body kinematics:

Theorem 6 The kinematic motor, Eq. (2.80), has in general a unique axis of rotation (α), called the central axis of rotation, such that all points along this axis move parallel to it and the resulting motion is a *twist* or *proper helicoidal motion.*

In analogy to Eq. (2.55) we get the vector equation for the central axis (α) of the motion,

$$\overrightarrow{\mathrm{O}_1\mathrm{O}_2} \times \boldsymbol{w} = \frac{1}{\omega^2}\left(\boldsymbol{w} \times \boldsymbol{v}^{\mathrm{O}_1}\right) \times \boldsymbol{w} \tag{2.82}$$

The general twisting motion of a rigid-body is given by the velocity of translation of a point O along the central axis, say $\boldsymbol{v}^{\mathrm{O}} = \boldsymbol{V}$, and the angular velocity vector $\boldsymbol{w}$ that defines in turn its axis of rotation. We use a coordinate system located at the considered point O and let x_i be the coordinate of any point P of the moving body. Then the velocity at point P is,

$$v_i^{\mathrm{P}} = V_i - \varepsilon_{ijk} x_j w_k \tag{2.83}$$

If we chose for example the translational velocity and the axis of rotation to be both vertical, then

$$\begin{aligned} V_i &= \delta_{i3} V \\ w_i &= \delta_{i3}\omega \end{aligned} \tag{2.84}$$

and with that,

$$\begin{aligned} v_1 &= -\varepsilon_{1j3} x_j \omega = -x_2\omega \\ v_2 &= -\varepsilon_{2j3} x_j \omega = +x_1\omega \\ v_3 &= V \end{aligned} \tag{2.85}$$

The position of particles attached to a normal, circular helix is,

$$\begin{aligned} x_1 &= a \cos\phi \\ x_2 &= a \sin\phi \\ x_3 &= b\phi \end{aligned} \tag{2.86}$$

and their velocity as the helix turns, becomes

$$v_i = \dot{x}_i \tag{2.87}$$

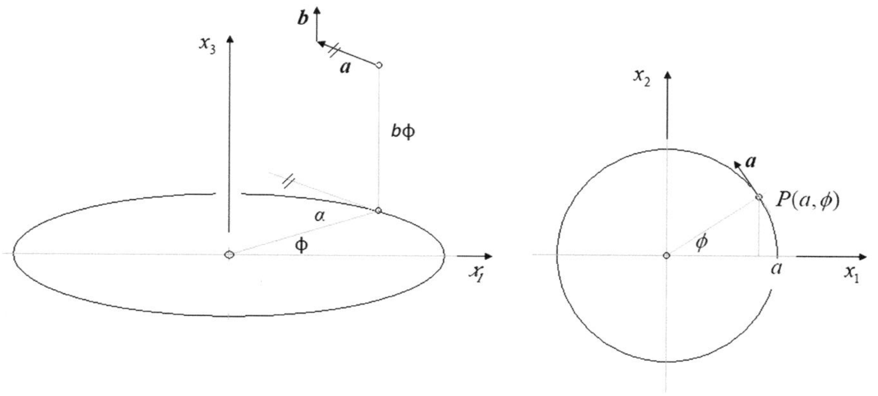

Fig. 2.18 Velocity of particles attached to the helix, Eq. (2.86)

where

$$\begin{aligned} v_1 &= \dot{x}_1 = -a \sin\phi\, \dot{\phi} = -a \sin\phi\, \omega \\ v_2 &= \dot{x}_2 = +a \cos\phi\, \dot{\phi} = +a \cos\phi\, \omega \\ v_3 &= \dot{x}_3 = b\dot{\phi} = b\omega \end{aligned} \tag{2.88}$$

The velocity is decomposed in a horizontal and a vertical component (Fig. 2.18),

$$\boldsymbol{v} = \boldsymbol{a} + \boldsymbol{b} \tag{2.89}$$

The component $\boldsymbol{a}$ is parallel to the $O(x_1, x_2)$—plane,

$$\boldsymbol{a} = v_1\boldsymbol{e}_1 + v_2\boldsymbol{e}_2 \tag{2.90}$$

and is tangential to the circle $O(a)$ at point $P(a, \phi)$, pointing in the direction of increasing ϕ; its modulus is

$$|\boldsymbol{a}| = a\omega \tag{2.91}$$

The axial component of the velocity is

$$\boldsymbol{b} = b\omega\boldsymbol{e}_3 \tag{2.92}$$

2.4 Motor Statics

2.4.1 Axiomatics

The statics of rigid bodies may be developed axiomatically and independently of dynamics [5]. The axioms of rigid body statics are elegantly presented by using the mathematical instrument of motor calculus.

Axiom 1 (transmissibility of force): A force acting on a rigid body may be shifted along its line of action (ε) so as to act on any point of that line.

$$\underline{F} = \begin{pmatrix} \boldsymbol{F} \\ \boldsymbol{M}^{\mathrm{A}} \end{pmatrix} = \begin{pmatrix} \boldsymbol{F} \\ \boldsymbol{M}^{\mathrm{B}} \end{pmatrix} \quad \forall \mathrm{A}, \mathrm{B} \in (\varepsilon) \tag{2.93}$$

since

$$\boldsymbol{M}^{\mathrm{B}} = \boldsymbol{M}^{\mathrm{A}} + \overrightarrow{\mathrm{AB}} \times \boldsymbol{F} = \boldsymbol{M}^{\mathrm{A}} \quad \forall \mathrm{A}, \mathrm{B} \in (\varepsilon) \tag{2.94}$$

Axiom 2 (addition of forces): Two forces acting on the same point A may be replaced by a single one, acting at this point and equal to their vector sum.

If,[6]

$$\underline{F} = \begin{pmatrix} \boldsymbol{F} \\ \boldsymbol{M}^{\mathrm{A}} \end{pmatrix}, \mathrm{A} \in (\varepsilon); \quad \boldsymbol{M}^{\mathrm{A}} = \mathbf{R}_{\mathrm{A}} \times \boldsymbol{F} \tag{2.95}$$

$$\underline{F}' = \begin{pmatrix} \boldsymbol{F}' \\ \boldsymbol{M}'^{\mathrm{A}} \end{pmatrix}, \mathrm{A} \in (\eta); \quad \boldsymbol{M}^{\mathrm{A}'} = \mathbf{R}_{\mathrm{A}} \times \boldsymbol{F}' \tag{2.96}$$

Then,

$$\underline{F} + \underline{F}' = \begin{pmatrix} \boldsymbol{F} + \boldsymbol{F}' \\ \boldsymbol{M}'^{\mathrm{A}} + \boldsymbol{M}'^{\mathrm{A}} \end{pmatrix}, \quad \mathrm{A} = (\varepsilon) \cap (\eta) \tag{2.97}$$

$$\boldsymbol{M}^{\mathrm{A}} + \boldsymbol{M}'^{\mathrm{A}} = \mathbf{R}_{\mathrm{A}} \times (\boldsymbol{F} + \boldsymbol{F}') \tag{2.98}$$

Rigid-body statics uses only Newton's 3rd law, thus:

Axiom 3 (action and reaction): Rigid bodies interact by pairs of opposed forces. In order to illustrate Axiom 3, let us consider the two rigid bodies, enumerated by the index $\alpha = 1, 2$ and denoted as (1) and (2), respectively (Fig. 2.19). According to Axiom 3, the two bodies interact with a pair of opposite forces $(\vec{F}^{\mathrm{A}}, -\vec{F}^{\mathrm{A}})$, acting along a line (ε), say realized at a point $\mathrm{A} \in (\varepsilon)$. We assume that $\vec{F}^{\mathrm{A}}$ is the force acted upon body (1) by body (2), and $-\vec{F}^{\mathrm{A}}$ its reaction, i.e. the force acted upon body (2) by body (1). Let P be a selected point,[7] say along the line that connects the centroids K_{α} of bodies (1) and (2). Transport of the interaction force-pair on point P, produces the following pair of proper motors,

$$\underline{F}^{(1)\mathrm{P}} = \begin{pmatrix} \boldsymbol{F}^{(1)\mathrm{P}} \\ \boldsymbol{M}^{(1)\mathrm{P}} \end{pmatrix}; \quad \boldsymbol{F}^{(1)\mathrm{P}} \cdot \boldsymbol{M}^{(1)\mathrm{P}} = 0 \tag{2.99}$$

---

[6]Note that in a rigorous but heavy presentation, in the expressions for the vector products one should use the point vector $\boldsymbol{f}$ instead of the free vector $\boldsymbol{F}$.

[7]The selection of the reduction point is arbitrary. However a general rule must be put down if one wants to produce some useful result.

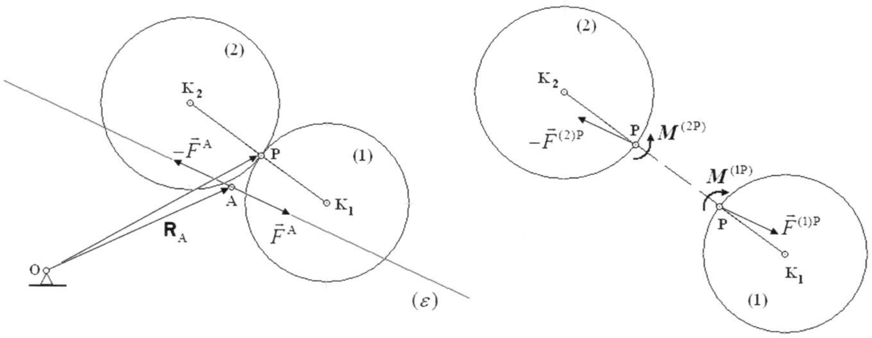

Fig. 2.19 Action and reaction between two rigid bodies

$$\underline{F}^{(2)\mathrm{P}} = \begin{pmatrix} \boldsymbol{F}^{(2)\mathrm{P}} \\ \boldsymbol{M}^{(2)\mathrm{P}} \end{pmatrix}; \quad \boldsymbol{F}^{(2)\mathrm{P}} \cdot \boldsymbol{M}^{(2)\mathrm{P}} = 0 \tag{2.100}$$

where

$$\begin{aligned} &\boldsymbol{F}^{(1)\mathrm{P}} = \boldsymbol{F} \\ &\boldsymbol{M}^{(1)\mathrm{P}} = \boldsymbol{M}^{(1)\mathrm{A}} + \overrightarrow{\mathrm{PA}} \times \boldsymbol{F}; \quad \boldsymbol{M}^{(1)\mathrm{A}} = \mathbf{R}_{\mathrm{A}} \times \boldsymbol{F} \end{aligned} \tag{2.101}$$

and

$$\begin{aligned} &\boldsymbol{F}^{(2)\mathrm{P}} = -\boldsymbol{F} \\ &\boldsymbol{M}^{(2)\mathrm{P}} = \boldsymbol{M}^{(2)\mathrm{A}} + \overrightarrow{\mathrm{PA}} \times (-\boldsymbol{F}); \quad \boldsymbol{M}^{(2)\mathrm{A}} = \mathbf{R}_{\mathrm{A}} \times (-\boldsymbol{F}) \end{aligned} \tag{2.102}$$

Thus

$$\boldsymbol{M}^{(2)\mathrm{P}} = -\boldsymbol{M}^{(1)\mathrm{P}} \tag{2.103}$$

and with that also Newton's 3rd law reads,

$$\underline{F}^{(2)\mathrm{P}} = -\underline{F}^{(1)\mathrm{P}} \tag{2.104}$$

The interaction force pair may by further transported to the centroids of the considered particles,

$$\underline{F}_{\mathrm{P}}^{\mathrm{K}_1} = \begin{pmatrix} \boldsymbol{F}_{\mathrm{P}}^{(1)} \\ \boldsymbol{M}_{\mathrm{P}}^{\mathrm{K}_1} \end{pmatrix} \tag{2.105}$$

$$\underline{F}_{\mathrm{P}}^{\mathrm{K}_2} = \begin{pmatrix} \boldsymbol{F}_{\mathrm{P}}^{(2)} \\ \boldsymbol{M}_{\mathrm{P}}^{\mathrm{K}_2} \end{pmatrix} \tag{2.106}$$

where

$$\boldsymbol{F}_{\mathrm{P}}^{(1)} = \boldsymbol{F}; \quad \boldsymbol{M}_{\mathrm{P}}^{\mathrm{K}_1} = \boldsymbol{M}^{(1)\mathrm{P}} + \overrightarrow{\mathrm{K}_1\mathrm{P}} \times \boldsymbol{F} \tag{2.107}$$

$$\boldsymbol{F}_{\mathrm{P}}^{(2)} = -\boldsymbol{F}; \quad \boldsymbol{M}_{\mathrm{P}}^{\mathrm{K}_2} = \boldsymbol{M}^{(2)\mathrm{P}} + \overrightarrow{\mathrm{K}_2\mathrm{P}} \times (-\boldsymbol{F}) \tag{2.108}$$

We note that the lines (ε) and $\mathrm{K}_1\mathrm{P}$ or $\mathrm{K}_2\mathrm{P}$ are in general skew. Thus, $\underline{F}_{\mathrm{P}}^{\mathrm{K}_1}$ and $\underline{F}_{\mathrm{P}}^{\mathrm{K}_2}$ are in general non-coaxial motors.

Assume now that a system of motors is acting on a rigid body. The resultant action is their motor sum,

$$\underline{F}^{\mathrm{K}} = \sum_p \underline{F}_p^{\mathrm{K}} = \begin{pmatrix} \boldsymbol{F} \\ \boldsymbol{M}^{\mathrm{K}} \end{pmatrix} \tag{2.109}$$

where

$$\begin{aligned} \boldsymbol{F} &= \sum_p \boldsymbol{F}_p \\ \boldsymbol{M}^{\mathrm{K}} &= \sum_p \boldsymbol{M}_p^{\mathrm{K}} \end{aligned} \tag{2.110}$$

This motor is in general equivalent to two skew line forces.

2.4.2 Equilibrium and Virtual Work Equation

Axiom 4 (static equilibrium): If the forces acting on a rigid body, initially at rest, can be reduced to zero by means of axioms 1 and 2, the body will remain at rest.

$$\underline{F}^{\mathrm{K}} = \underline{0} \Leftrightarrow \quad \boldsymbol{F} = \sum_p \boldsymbol{F}_p = \boldsymbol{0} \ \wedge \ \boldsymbol{M}^{\mathrm{K}} = \sum_p \boldsymbol{M}_p^{\mathrm{K}} = \boldsymbol{0} \tag{2.111}$$

Let $\delta\underline{\kappa}^{\mathrm{K}}$ be the kinematic motor for a virtual displacement of the considered rigid body,

$$\delta\bar{\kappa}^{\mathrm{K}} = \begin{pmatrix} \delta\boldsymbol{w} \\ \delta\boldsymbol{v}^{\mathrm{K}} \end{pmatrix} \tag{2.112}$$

where $\delta\boldsymbol{w}$ and $\delta\boldsymbol{v}^{\mathrm{K}}$ are the corresponding virtual spin vector and virtual velocity vector of the centroid K, respectively.

The virtual power of the force and couple are defined as the corresponding *von Mises motor scalar product* [2, 3],

$$\delta W = \underline{F}^{\mathrm{K}} \circ \delta \bar{\kappa} = \begin{pmatrix} \boldsymbol{F} \\ \boldsymbol{M}^{\mathrm{K}} \end{pmatrix} \circ \begin{pmatrix} \delta \boldsymbol{w} \\ \delta \boldsymbol{v}^{\mathrm{K}} \end{pmatrix} = \boldsymbol{F} \cdot \delta \boldsymbol{v}^{\mathrm{K}} + \boldsymbol{M}^{\mathrm{K}} \cdot \delta \boldsymbol{w} \tag{2.113}$$

Let us consider the case where instead of point K a different reduction point was chosen, say K′. From Eqs. (2.44) and (2.79) we get,

$$\boldsymbol{M}^{\mathrm{K}} = \boldsymbol{M}^{\mathrm{K}'} + \overrightarrow{\mathrm{KK}'} \times \boldsymbol{F} \tag{2.114}$$

$$\delta \boldsymbol{v}^{\mathrm{K}} = \delta \boldsymbol{v}^{\mathrm{K}'} + \overrightarrow{\mathrm{KK}'} \times \delta \boldsymbol{w} \tag{2.115}$$

From these expressions and Eq. (2.113) we obtain that,

$$\begin{aligned} \delta W &= \boldsymbol{F} \cdot \left(\delta \boldsymbol{v}^{\mathrm{K}'} + \overrightarrow{\mathrm{KK}'} \times \delta \boldsymbol{w} \right) + \left(\boldsymbol{M}^{\mathrm{K}'} + \overrightarrow{\mathrm{KK}'} \times \boldsymbol{F} \right) \cdot \delta \boldsymbol{w} \\ &= \boldsymbol{F} \cdot \delta \boldsymbol{v}^{\mathrm{K}'} + \boldsymbol{F} \cdot \left(\overrightarrow{\mathrm{KK}'} \times \delta \boldsymbol{w} \right) + \boldsymbol{M}^{\mathrm{K}'} \cdot \delta \boldsymbol{w} + \left(\overrightarrow{\mathrm{KK}'} \times \boldsymbol{F} \right) \cdot \delta \boldsymbol{w} \end{aligned} \tag{2.116}$$

Since

$$\boldsymbol{F} \cdot \left(\overrightarrow{\mathrm{KK}'} \times \delta \boldsymbol{w} \right) = -\left(\overrightarrow{\mathrm{KK}'} \times \boldsymbol{F} \right) \cdot \delta \boldsymbol{w} \tag{2.117}$$

it follows that the value of the von Mises motor scalar product is invariant with respect to changes in the position of the reduction point,

$$\delta W = \underline{F}^{\mathrm{K}} \circ \delta \bar{\kappa}^{\mathrm{K}} = \underline{F}^{\mathrm{K}'} \circ \delta \bar{\kappa}^{\mathrm{K}'} \tag{2.118}$$

Finally, we observe that from the virtual work equation,

$$\delta W = 0 \tag{2.119}$$

and for independent variation of $\delta \boldsymbol{w}$ and $\delta \boldsymbol{v}^{\mathrm{K}}$ we get the equilibrium Eq. (2.111) and conversely from the equilibrium Eq. (2.111) we get the virtual power Eq. (2.119).

References

1. Clifford, W. K. (1873). Preliminary sketch of bi-quaternions. *Proceedings of the London Mathematical Society, 4,* 381–395.
2. Mises, R. V. (1924a). Motorrechnung, ein neues Hilfsmittel der Mechanik. *Zeitschrift für Angewandte Mathematik und Mechanik,* **4**, 155–181.
3. Mises, R. V. (1924b) Anwendungen der Motorrechnung. *Zeitschrift für Angewandte Mathematik und Mechanik,* **4**, 193–213.

4. Study, E. (1903). *Geometrie der Dynamen. Die Zusammensetzung von Kräften und verwandte Gegenstände*. Teubner.
5. Brand, L. (1940). *Vector and tensor analysis*. Wiley.
6. Talpaert, Y. R. (2003). *Mechanics, tensors and virtual works*. Cambridge International Science Publishing.
7. Chasles, M. (1830). Note sur les propriétés générales du système de deux corps semblables entr'eux. *Bulletin de Sciences Mathématiques, Astronomiques Physiques et Chimiques, Baron de Ferussac, Paris, 14,* 321–326.
8. Ceccarelli, M. (2000). Screw axis defined by Giulio Mozzi in 1763 and early studies on helicoidal motion. *Mechanism and Machine Theory, 35,* 761–770.
9. Mozzi, G. (1763). *Discorso matematico sopra il rotamento momentaneo dei corpi*. Napoli: Stamperia di Donato Campo.
10. Ball, R. S. (1900). *A treatise on the theory of screws*. Cambridge University Press.

Chapter 3
Cosserat Continuum Kinematics

Abstract This chapter derives the kinematic fields (deformation and deformation rate tensors) in general curvilinear coordinates, before reducing them to the familiar forms of Cosserat continuum in Cartesian coordinates. It showcases this way that the motor calculus approach has the same information as the classical representation as a limiting case, but can be used in a generic framework. It finishes with the integrability, compatibility and discontinuity conditions for the considered representation.

3.1 Motion and Deformation in General Coordinates

Let the position vector of a point in the three dimensional space be denoted as (Fig. 3.1),

$$\vec{OP} = \mathbf{R} = x^i \boldsymbol{e}_i \tag{3.1}$$

where x^i and $\boldsymbol{e}_i$ are the underlying Cartesian coordinates of the position vector and the Cartesian basis vectors, respectively. Let $\Theta^i (i = 1, 2, 3)$ denote fixed in space general curvilinear coordinates, that are related to the Cartesian coordinates through the transformation,

$$x^i = \chi^i(\Theta^k); \quad \left| \frac{\partial \chi^i}{\partial \Theta^k} \right| \neq 0 \tag{3.2}$$

This transformation allows us to write the position vector as a function of the curvilinear coordinates Θ^i of the point P

$$\mathbf{R} = \mathbf{R}(\Theta^i) \tag{3.3}$$

and to introduce at any point in space the local covariant affine basis

$$\boldsymbol{g}_i = \frac{\partial \mathbf{R}}{\partial \Theta^i} = \mathbf{R}_{,i}, \quad (\cdot)_{,i} \equiv \frac{\partial}{\partial \Theta^i} \tag{3.4}$$

I. Vardoulakis, *Cosserat Continuum Mechanics*, Lecture Notes in Applied and Computational Mechanics 87, https://doi.org/10.1007/978-3-319-95156-0_3

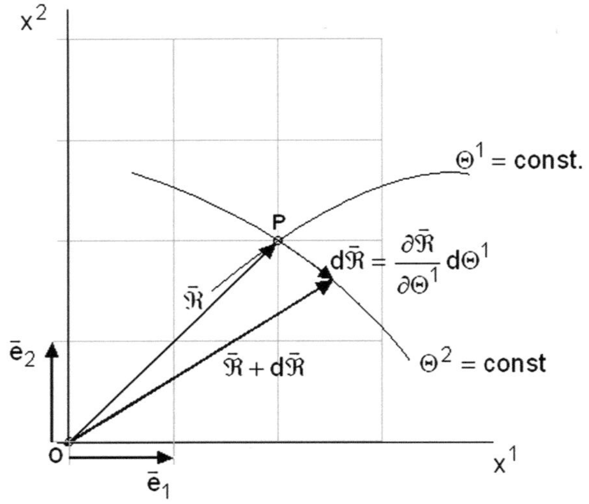

Fig. 3.1 Cartesian and curvilinear coordinates of point in the plane

We assume that the basis vectors, $\boldsymbol{g}_1, \boldsymbol{g}_2, \boldsymbol{g}_3$, in the given order build a right-handed system.

Let a Cosserat continuum particle be located at point $P(\Theta^i)$. This particle is seen as a rigid body of infinitesimal dimensions and has the degrees of freedom of rigid-body displacement and rigid-body rotation. We restrict here our analysis to infinitesimal motions. The infinitesimal particle rotation is an axial vector and we emphasize that this statement is not true for finite rotations, that are not considered here.

The motion of the Cosserat particle is described by its rotation vector, that is in turn described primarily by its contravariant components (Fig. 3.2),

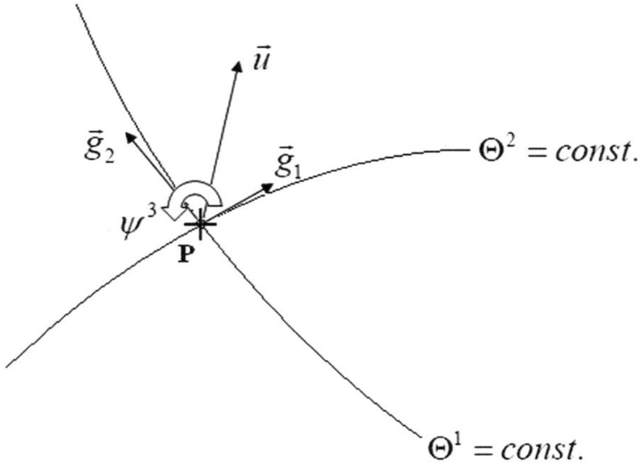

Fig. 3.2 Dofs of a 2D-Cosserat particle and the local affine covariant basis vectors $(\boldsymbol{g}_1, \boldsymbol{g}_2)$

$$\boldsymbol{\psi} = \psi^i(\Theta^k)\boldsymbol{g}_i \tag{3.5}$$

and its displacement vector, that is described by its covariant components,

$$\boldsymbol{u} = u_i(\Theta^k)\boldsymbol{g}^i \tag{3.6}$$

where $\boldsymbol{g}_i$ and $\boldsymbol{g}^i$ are the covariant and contravariant bases respectively at point $P(\Theta^i)$.

For the compact description of the Cosserat continuum particle kinematics we follow Schaefer's suggestion [1] and we introduce the vector compound,

$$\bar{K}(\Theta^i) = \begin{pmatrix} \boldsymbol{\psi}(\Theta^i) \\ \boldsymbol{u}(\Theta^i) \end{pmatrix} = \begin{pmatrix} \psi^i \boldsymbol{g}_i \\ u_i \boldsymbol{g}^i \end{pmatrix} \tag{3.7}$$

It is important to note that the first entry in vector compound, Eq. (3.7), is a contravariant vector and the second a covariant vector. This distinction will be lost if we use Cartesian coordinates and we get the false impression that both entries are of the same nature.

If the Cosserat continuum in the vicinity of a point $P(\Theta^k)$ is not deforming, then the motion in this neighbourhood is that of a rigid body. In this case we have that,

$$\boldsymbol{\psi}(\Theta^i + d\Theta^i) = \boldsymbol{\psi}(\Theta^i) \quad \Rightarrow \quad d\boldsymbol{\psi} = \boldsymbol{0} \tag{3.8}$$

since in rigid-body mechanics the infinitesimal rotation vector is independent of the point of reference. On the other hand, the infinitesimal displacement vector obeys the transport law of rigid-body kinematics,

$$\boldsymbol{u}(\Theta^i + d\Theta^i) = \boldsymbol{u}(\Theta^i) + \boldsymbol{\psi}(\Theta^i) \times d\Theta^k \boldsymbol{g}_k \quad \Rightarrow \quad d\boldsymbol{u} = \boldsymbol{\psi}(\Theta^i) \times d\Theta^k \boldsymbol{g}_k \tag{3.9}$$

Thus, the kinematic compound in the infinitesimal neighborhood of point $P(\Theta^k)$ is,

$$\bar{K}' = \bar{K}(\Theta^i + d\Theta^i) = \begin{pmatrix} \boldsymbol{\psi}(\Theta^i) \\ \boldsymbol{u}(\Theta^i) + \boldsymbol{\psi}(\Theta^i) \times d\Theta^l \boldsymbol{g}_l \end{pmatrix} = \begin{pmatrix} \psi^i \boldsymbol{g}_i \\ (u_i + e_{ikl}\psi^k d\Theta^l)\boldsymbol{g}^i \end{pmatrix} \tag{3.10}$$

where e_{klm} is the corresponding Levi-Civita 3rd-order fully antisymmetric tensor,

$$e_{klm} = \begin{cases} \sqrt{g} & if : (k,l,m) = cycl(1,2,3) \\ -\sqrt{g} & if : (k,l,m) = cycl(2,1,3)\,, \\ 0 & else \end{cases} \quad g = \det(g_{ij}) > 0 \tag{3.11}$$

and g_{ij} is the covariant metric tensor, associated to the chosen covariant basis.

With,

$$\bar{K}' = \bar{K}(\Theta^i + d\Theta^i) = \begin{pmatrix} \psi'^i \boldsymbol{g}_i \\ u'_i \boldsymbol{g}^i \end{pmatrix} \tag{3.12}$$

from Eq. (3.10) follows that the components of the two vector compounds $\bar{K}(\Theta^i)$ and $\bar{K}(\Theta^i + d\Theta^i)$ are related as,

$$\begin{aligned} \psi'^i &= \psi^i \\ u'_i &= u_i + e_{ikl}\psi^k d\Theta^l \end{aligned} \tag{3.13}$$

This means that in case of a locally non-deforming continuum the above introduced compound $\bar{K}$ of the two vectors $\boldsymbol{\psi}$ and $\boldsymbol{u}$, Eq. (3.7), is a motor in the sense of von Mises. From Eq. (3.13) follows that the two motors $\bar{K}$ and $\bar{K}'$ are "*equal*" or "*kinematically equivalent*". This fundamental kinematical property of Cosserat continua has motivated their application to the mechanics of granular media. In that case the single, rigid grain is seen as the smallest material unit.

In general the differential forms,

$$d\boldsymbol{\psi} = \boldsymbol{\psi}(\Theta^i + d\Theta^i) - \boldsymbol{\psi}(\Theta^i) \tag{3.14}$$

$$d\boldsymbol{u} - \boldsymbol{\psi}(\Theta^i) \times d\Theta^k \boldsymbol{g}_k = \boldsymbol{u}(\Theta^i + d\Theta^i) - (\boldsymbol{u}(\Theta^i) + \boldsymbol{\psi}(\Theta^i) d\Theta^k \boldsymbol{g}_k) \tag{3.15}$$

will not vanish. In this case we will say that the neighborhood of the point $P(\Theta^k)$ is deforming. In compact form the deformation is described by the absolute differential,

$$d\bar{K}(\Theta^i) = \bar{K}(\Theta^i + d\Theta^i) - \bar{K}(\Theta^i) \tag{3.16}$$

or by the differential compound,

$$d\bar{K}(\Theta^i) = \begin{pmatrix} \boldsymbol{\psi}(\Theta^i + d\Theta^i) - \boldsymbol{\psi}(\Theta^i) \\ \boldsymbol{u}(\Theta^i + d\Theta^i) - (\boldsymbol{u}(\Theta^i) + \boldsymbol{\psi}(\Theta^i) \times d\Theta^k \boldsymbol{g}_k) \end{pmatrix} = \begin{pmatrix} d\boldsymbol{\psi} \\ d\boldsymbol{u} + d\Theta^k \boldsymbol{g}_k \times \boldsymbol{\psi} \end{pmatrix} \tag{3.17}$$

In view of Eqs. (3.14), (3.15) and (3.17) we introduce the Pfaffian vector forms

$$\boldsymbol{\kappa}_i d\Theta^i = d\boldsymbol{\psi} \tag{3.18}$$

$$\boldsymbol{\gamma}_i d\Theta^i = d\boldsymbol{u} + d\Theta^k \boldsymbol{g}_k \times \boldsymbol{\psi} \tag{3.19}$$

These forms define in turn the two vectors,

$$\boldsymbol{\kappa}_i = \boldsymbol{\psi}_{,i} \tag{3.20}$$

and

$$\gamma_i = \boldsymbol{u}_{,i} + \boldsymbol{g}_i \times \boldsymbol{\psi} \tag{3.21}$$

We recall that the gradient of a vector is expressed by means of its covariant derivative,

$$\boldsymbol{\psi}_{,i} = \psi^{k}_{\cdot\, i}\boldsymbol{g}_k \tag{3.22}$$

and

$$\boldsymbol{u}_{,i} = u_{k|i}\boldsymbol{g}^k \tag{3.23}$$

where $()_{i|j}$ denotes covariant differentiation of a vector; e.g.,

$$a_{i|j} = a_{i,j} - \Gamma^k_{ij}a_k \tag{3.24}$$

and

$$\Gamma^i_{jk} \equiv \left\{ \begin{matrix} i \\ jk \end{matrix} \right\} \tag{3.25}$$

are the Christoffel symbols of the second kind.

We notice also that the ith component of the vector product of two vectors is computed

$$(\boldsymbol{x} \times \boldsymbol{y})_i = e_{ikl}x^k y^l \tag{3.26}$$

Thus, the 2nd term on r.h.s. of Eq. (3.21) becomes,

$$\boldsymbol{g}_i \times \boldsymbol{\psi} = \psi^k \boldsymbol{g}_i \times \boldsymbol{g}_k = \psi^k e_{ilk}\boldsymbol{g}^l = -e_{ilk}\psi^k \boldsymbol{g}^l \tag{3.27}$$

From Eqs. (3.20) and (3.22) we get that,

$$\boldsymbol{\kappa}_i = \psi^{k}_{\cdot\,|i}\boldsymbol{g}_k \tag{3.28}$$

and from Eqs. (3.21), (3.23) and (3.27) we get that

$$\gamma_i = (u_{k|i} - e_{ikl}\psi^l)\boldsymbol{g}^k \tag{3.29}$$

With

$$\boldsymbol{\kappa}_i = \kappa_{i}^{\cdot\,k} \boldsymbol{g}_k \tag{3.30}$$

and

$$\boldsymbol{\gamma}_i = \gamma_{ik} \boldsymbol{g}^k \tag{3.31}$$

The components of the above introduced deformation vectors define the corresponding deformation tensors,

$$\kappa_i^{\cdot k} = \psi^k_{\cdot\,|i} \tag{3.32}$$

$$\gamma_{ik} = u_{k|i} - e_{ikl}\psi^l \tag{3.33}$$

With

$$d\bar{K}(\Theta^i) = \begin{pmatrix} \boldsymbol{\kappa}_i \\ \boldsymbol{\gamma}_i \end{pmatrix} d\Theta^i = \begin{pmatrix} \kappa_{i}^{\cdot\,k} \boldsymbol{g}_k \\ \gamma_{ik}\boldsymbol{g}^k \end{pmatrix} d\Theta^i = \begin{pmatrix} \psi^k_{\cdot\,|i}\boldsymbol{g}_k \\ (u_{k|i} + e_{kil}\psi^l)\boldsymbol{g}^k \end{pmatrix} d\Theta^i \tag{3.34}$$

the deformation that is induced by the vector fields, Eqs. (3.5) and (3.6), is best illustrated if we introduce the tensor compound,

$$\bar{\bar{H}} = \begin{pmatrix} \kappa_{i}^{\cdot\,k} \boldsymbol{g}^i \otimes \boldsymbol{g}_k \\ \gamma_{ik}\boldsymbol{g}^i \otimes \boldsymbol{g}^k \end{pmatrix} = \begin{pmatrix} \psi^k_{\cdot|i}\boldsymbol{g}_k \otimes \boldsymbol{g}^i \\ (u_{k|i} + e_{kil}\psi^l)\boldsymbol{g}^k \otimes \boldsymbol{g}^i \end{pmatrix} \tag{3.35}$$

In Cosserat continuum mechanics the compound $\bar{\bar{H}}$ plays the role of a generalized-displacement gradient.

Following Kessel [2], the above observations prompt the definition of a gradient operator that is applied onto the kinematic motor $\bar{K}$ and produces the generalized-displacement gradient,

$$\bar{K} = \begin{pmatrix} \psi^i \boldsymbol{g}_i \\ u_i \boldsymbol{g}^i \end{pmatrix} \quad \Rightarrow \quad \bar{\bar{H}} = Grad\bar{\mathrm{K}} := \begin{pmatrix} \psi^i_{\cdot|k}\boldsymbol{g}_i \otimes \boldsymbol{g}^k \\ (u_{i|k} + e_{ikl}\psi^l)\boldsymbol{g}^i \otimes \boldsymbol{g}^k \end{pmatrix} \tag{3.36}$$

Thus from the 6 placements ψ^i and u_i $(i = 1, 2, 3)$ we have generated 18 deformations $\kappa_i^{\cdot k}$ and γ_{ik}. The tensor $\kappa_{i}^{\cdot\,k}$, Eq. (3.32), is called the infinitesimal tensor of distortions; the components $\kappa_{(i)}^{\cdot\,(i)}$ are called infinitesimal "torsions" and the rest components are the infinitesimal "curvatures". We call γ_{ik}, Eq. (3.33), the infinitesimal (relative) deformation tensor. Its symmetric part coincides with the usual infinitesimal strain tensor,

$$\gamma_{(ik)} = \frac{1}{2}(\gamma_{ik} + \gamma_{ki}) = \varepsilon_{ij} \tag{3.37}$$

where

$$\varepsilon_{ij} = \frac{1}{2}\left(u_{k|i} + u_{i|k}\right) \tag{3.38}$$

The antisymmetric part of γ_{ik} coincides with the relative rotation,

$$\gamma_{[ik]} = \frac{1}{2}(\gamma_{ik} - \gamma_{ki}) = \psi_{ik} - \omega_{ik} \tag{3.39}$$

where

$$\omega_{ik} = \frac{1}{2}\left(u_{i|k} - u_{k|i}\right) \tag{3.40}$$

and

$$\psi_{ij} = -e_{ijk}\psi^k \tag{3.41}$$

Let ω^k be the axial vector that corresponds to ω_{ij},

$$\omega_{ij} = -e_{ijk}\omega^k \tag{3.42}$$

Then the antisymmetric part of γ_{ik} is indeed given in terms of the difference of the two related axial vectors,

$$\gamma_{[ij]} = e_{ijk}\left(\omega^k - \psi^k\right) \tag{3.43}$$

This property justifies the name *relative deformation tensor* that is given to γ_{ik}.

3.2 Cosserat Kinematics in Cartesian Coordinates

3.2.1 Strain, Spin, Curvature and Torsion

Let x_i be the Cartesian coordinates of a point of a rigid body before the motion and x_i' the coordinates of the same point after the motion. We consider two neighbouring points $P(x_i)$ and $Q(y_i)$ in the undeformed configuration of a Cosserat continuum, such that $y_i = x_i + dx_i$. The material line element that connects these two points is given by the vector,

$$\vec{PQ} = dx_i \boldsymbol{e}_i \tag{3.44}$$

The positions of points P and Q in the deformed configuration are computed as,

$$\begin{aligned} x'_i &= x_i + u_i \\ y'_i &= x_i + dx_i + u_i + \partial_j u_i dx_j \end{aligned} \tag{3.45}$$

where ∂_i is the Cartesian differentiation operator,

$$\partial_i \equiv \frac{\partial}{\partial x_i} \tag{3.46}$$

Thus

$$\begin{aligned} dx'_i = y'_i - x'_i &= x_i + dx_i + u_i + \partial_j u_i dx_j - (x_i + u_i) \\ &= dx_i + \partial_j u_i dx_j = \left(\delta_{ij} + \partial_j u_i\right) dx_j \end{aligned} \tag{3.47}$$

The length of the line element before and after the deformation is

$$\begin{aligned} ds^2 &= dx_i dx_i \\ ds'^2 &= dx'_i dx'_i = \left(\delta_{ij} + \partial_j u_i\right) dx_j (\delta_{ik} + \partial_k u_i) dx_k \\ &= \delta_{kj} dx_j dx_k + \left(\partial_k u_j + \partial_j u_k + \partial_j u_i \partial_k u_i\right) dx_j dx_k \\ &\approx ds^2 + 2\gamma_{(ij)} dx_i dx_j \end{aligned} \tag{3.48}$$

where $\gamma_{(ij)}$ is the symmetric part of the relative deformation tensor,

$$\gamma_{(ij)} = \frac{1}{2}\left(\gamma_{ij} + \gamma_{ji}\right) = \frac{1}{2}\left(\partial_i u_j + \partial_j u_i\right) \tag{3.49}$$

that coincides in turn with common infinitesimal strain tensor in the Boltzmann continuum,

$$\gamma_{(ij)} = \varepsilon_{ij} \tag{3.50}$$

where

$$\varepsilon_{ij} = \frac{1}{2}\left(\partial_i u_j + \partial_j u_i\right) \tag{3.51}$$

Let us now consider the antisymmetric part of the relative deformation tensor,

$$\gamma_{[ij]} = \frac{1}{2}\left(\gamma_{ij} - \gamma_{ji}\right) = \frac{1}{2}\left(\partial_i u_j - \partial_j u_i\right) - \varepsilon_{ijk}\psi_k \tag{3.52}$$

The antisymmetric part of the transposed displacement gradient is denoted as

$$\omega_{ij} = \frac{1}{2}\left(\partial_j u_i - \partial_i u_j\right) \tag{3.53}$$

Thus

$$\gamma_{[ij]} = \psi_{ij} - \omega_{ij} \tag{3.54}$$

where

$$\psi_{ij} = -\varepsilon_{ijk}\psi_k \tag{3.55}$$

We may define the axial vector ω_k that corresponds to ω_{ij},

$$\omega_k = -\frac{1}{2}\varepsilon_{ijk}\omega_{ij} \quad \Leftrightarrow \quad \omega_{ij} = -\varepsilon_{ijk}\omega_k \tag{3.56}$$

with

$$\omega_k = \omega\mu_k; \quad \mu_k\mu_k = 1 \tag{3.57}$$

With this notation Eq. (3.56) becomes,

$$\gamma_{[ij]} = \psi_{ij} - \omega_{ij} = -\varepsilon_{ijk}\psi_k + \varepsilon_{ijk}\omega_k = -\varepsilon_{ijk}(\psi_k - \omega_k) \tag{3.58}$$

Summarizing the above results, we get from Eqs. (3.49) to (3.58),

$$\gamma_{ij} = \varepsilon_{ij} + \left(\psi_{ij} - \omega_{ij}\right) = \varepsilon_{ij} - \varepsilon_{ijk}(\psi_k - \omega_k) \tag{3.59}$$

From the above equation we get,

$$\gamma_{ji} + \varepsilon_{jik}\psi_k = \varepsilon_{ji} + \varepsilon_{jik}\omega_k \tag{3.60}$$

In Cartesian components, from Eq. (3.33) we get

$$\partial_j u_i = \gamma_{ji} + \varepsilon_{jik}\psi^k \tag{3.61}$$

From Eqs. (3.47) we get,

$$\Delta dx_i = dx'_i - dx_i = \partial_j u_i dx_j = \left(\gamma_{ji} + \varepsilon_{jik}\psi_k\right)dx_j \tag{3.62}$$

and with Eq. (3.60) it reduces to the familiar linear form

$$\Delta dx_i = \left(\varepsilon_{ji} + \varepsilon_{jik}\omega_k\right)dx_j \tag{3.63}$$

Remark
In matrix notation Eq. (3.56) reads as,

$$\left[\omega_{ij}\right] = \begin{bmatrix} 0 & \omega_{12} & \omega_{13} \\ \omega_{21} & 0 & \omega_{23} \\ \omega_{31} & \omega_{32} & 0 \end{bmatrix} = \begin{bmatrix} 0 & -\varepsilon_{123}\omega_3 & -\varepsilon_{132}\omega_2 \\ -\varepsilon_{213}\omega_3 & 0 & -\varepsilon_{231}\omega_1 \\ -\varepsilon_{312}\omega_2 & -\varepsilon_{321}\omega_1 & 0 \end{bmatrix}$$
$$= \begin{bmatrix} 0 & -\omega_3 & \omega_2 \\ \omega_3 & 0 & -\omega_1 \\ -\omega_2 & \omega_1 & 0 \end{bmatrix} \tag{3.64}$$

Or in indicial notation

$$\omega_m = \frac{1}{2}\varepsilon_{mkl}\partial_k u_l \tag{3.65}$$

Explicitly, in components we have,

$$\omega_1 = \frac{1}{2}\varepsilon_{1kl}\partial_k u_l = \frac{1}{2}(\varepsilon_{123}\partial_2 u_3 + \varepsilon_{132}\partial_3 u_2) = \frac{1}{2}(\partial_2 u_3 - \partial_3 u_2) = \omega_{32} \tag{3.66}$$

$$\begin{aligned} \omega_2 &= \frac{1}{2}\varepsilon_{2kl}\partial_k u_l = \frac{1}{2}(\varepsilon_{231}\partial_3 u_1 + \varepsilon_{213}\partial_1 u_3)\frac{1}{2}(\partial_3 u_1 - \partial_1 u_3) = \omega_{13} \\ \omega_3 &= \frac{1}{2}\varepsilon_{3kl}\partial_k u_l = \frac{1}{2}(\varepsilon_{312}\partial_1 u_2 + \varepsilon_{321}\partial_2 u_1) = \frac{1}{2}(\partial_1 u_2 - \partial_2 u_1) = \omega_{21} \end{aligned} \tag{3.67}$$

With,

$$\mathbf{rot}\,\boldsymbol{u} = \left(\frac{\partial u_3}{\partial x_2} - \frac{\partial u_2}{\partial x_3}\right)\boldsymbol{e}_1 + \left(\frac{\partial u_1}{\partial x_3} - \frac{\partial u_3}{\partial x_1}\right)\boldsymbol{e}_2 + \left(\frac{\partial u_2}{\partial x_1} - \frac{\partial u_1}{\partial x_2}\right)\boldsymbol{e}_3 \tag{3.68}$$

we retrieve the well-known result,

$$\boldsymbol{\omega} = \omega_i \boldsymbol{e}_i = \frac{1}{2}\mathbf{rot}\,\boldsymbol{u} \tag{3.69}$$

Note that,

$$\begin{aligned} &\partial_m\omega_m = \frac{1}{2}\varepsilon_{mkl}\partial_m\partial_k u_l = \frac{1}{2}\varepsilon_{kml}\partial_k\partial_m u_l \quad \Rightarrow \quad \frac{1}{2}\varepsilon_{mkl}\partial_m\partial_k u_l - \frac{1}{2}\varepsilon_{kml}\partial_k\partial_m u_l = 0 \\ &\Rightarrow \frac{1}{2}\varepsilon_{mkl}\partial_m\partial_k u_l + \frac{1}{2}\varepsilon_{mkl}\partial_k\partial_m u_l = 0 \quad \Rightarrow \quad \varepsilon_{mkl}\partial_m\partial_k u_l = 0 \quad \Rightarrow \quad \partial_m\omega_m = 0 \end{aligned} \tag{3.70}$$

or symbolically,

$$\mathrm{div}\boldsymbol{\omega} = 0 \tag{3.71}$$

Equation (3.71) follows directly from Eq. (3.69), since for any vector $\boldsymbol{u}$ holds the identity,

$$\mathrm{div}\,\mathbf{rot}\,\boldsymbol{u} \equiv 0 \tag{3.72}$$

Note also that in general the divergence of the mean torsion is non-zero,

$$\mathrm{div}\boldsymbol{\psi} = \partial_m \psi_m = \kappa_{mm} \neq 0 \tag{3.73}$$

3.2.2 2D Cosserat Kinematics

As an application we assume a 2D setting. In this case we have the following placements [3],

$$\begin{aligned} \boldsymbol{u} &= u_1 \boldsymbol{e}_1 + u_2 \boldsymbol{e}_2 \\ \boldsymbol{\psi} &= \psi_3 \boldsymbol{e}_3 \end{aligned} \tag{3.74}$$

The components of the relative deformation tensor are

$$\begin{aligned} \gamma_{11} &= \partial_1 u_1 \\ \gamma_{12} &= \partial_1 u_2 + \psi_{12} = \partial_1 u_2 - \varepsilon_{123}\psi_3 = \partial_1 u_2 - \psi \\ \gamma_{21} &= \partial_2 u_1 + \psi_{21} = \partial_2 u_1 - \varepsilon_{213}\psi_3 = \partial_2 u_1 + \psi \\ \gamma_{22} &= \partial_2 u_2 \end{aligned} \tag{3.75}$$

Similarly, the components of the curvature tensor become,

$$\begin{aligned} \kappa_{11} &= \kappa_{12} = 0; \quad \kappa_{13} = \partial_1 \psi_3 = \partial_1 \psi \\ \kappa_{21} &= \kappa_{22} = 0; \quad \kappa_{23} = \partial_2 \psi_3 = \partial_2 \psi \\ \kappa_{11} &= \kappa_{22} = \kappa_{33} = 0 \end{aligned} \tag{3.76}$$

Introducing the infinitesimal strain tensor,

$$\begin{aligned} \varepsilon_{11} &= \partial_1 u_1 \\ \varepsilon_{12} &= \varepsilon_{21} = \frac{1}{2}(\partial_1 u_2 + \partial_2 u_1) \\ \varepsilon_{22} &= \partial_2 u_2 \end{aligned} \tag{3.77}$$

and the infinitesimal background rotation tensor

$$\begin{aligned}\omega_{11} &= 0\\ \omega_{12} &= \frac{1}{2}(\partial_2 u_1 - \partial_1 u_2) = -\varepsilon_{123}\omega_3 = -\omega\\ \omega_{21} &= \frac{1}{2}(\partial_1 u_2 - \partial_2 u_1) = -\varepsilon_{213}\omega_3 = +\omega\\ \omega_{22} &= 0\end{aligned} \tag{3.78}$$

we get,

$$\begin{aligned}\gamma_{11} &= \varepsilon_{11}\\ \gamma_{12} &= \partial_1 u_2 \pm \frac{1}{2}\partial_2 u_1 - \psi = \varepsilon_{12} + (\omega - \psi)\\ \gamma_{21} &= \partial_2 u_1 \pm \frac{1}{2}\partial_1 u_2 + \psi = \varepsilon_{21} - (\omega - \psi)\\ \gamma_{22} &= \partial_2 u_2\end{aligned} \tag{3.79}$$

Thus

$$\gamma_{(ij)} = \varepsilon_{ij} \tag{3.80}$$

and

$$\begin{aligned}\gamma_{[12]} &= \frac{1}{2}(\gamma_{12} - \gamma_{21}) = \frac{1}{2}(\partial_1 u_2 - \psi - (\partial_2 u_1 + \psi))\\ &= \frac{1}{2}(\partial_1 u_2 - \partial_2 u_1) - \psi = \omega - \psi\end{aligned} \tag{3.81}$$

We consider a line element $\vec{PQ}$ that is originally parallel to the x_1-axis, Eq. (3.44); Fig. 3.3. With

$$\begin{Bmatrix} dx_1 \\ dx_2 \end{Bmatrix} = \begin{Bmatrix} 1 \\ 0 \end{Bmatrix} dx \tag{3.82}$$

we get from Eq. (3.63),

$$\begin{Bmatrix} \Delta dx_1 \\ \Delta dx_2 \end{Bmatrix} = \begin{bmatrix} \varepsilon_{11} & \varepsilon_{21} - \omega \\ \varepsilon_{12} + \omega & \omega_{22} \end{bmatrix} \begin{Bmatrix} 1 \\ 0 \end{Bmatrix} dx = \begin{Bmatrix} \varepsilon_{11} dx \\ (\varepsilon_{12} + \omega) dx \end{Bmatrix} \tag{3.83}$$

Similarly for a line element $\vec{PR}$ that is originally parallel to the x_2-axis we get,

$$\begin{Bmatrix} \Delta dx_1 \\ \Delta dx_2 \end{Bmatrix} = \begin{bmatrix} \varepsilon_{11} & \varepsilon_{21} - \omega \\ \varepsilon_{12} + \omega & \omega_{22} \end{bmatrix} \begin{Bmatrix} 0 \\ 1 \end{Bmatrix} dy = \begin{Bmatrix} (\varepsilon_{21} - \omega) dy \\ \varepsilon_{22} dy \end{Bmatrix} \tag{3.84}$$

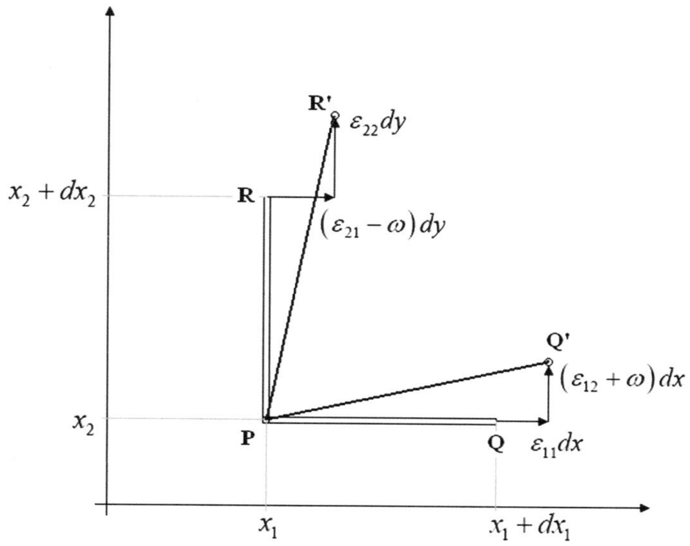

Fig. 3.3 The deformation of a solid orthogonal element

In Fig. 3.3 we show the geometrical visualization of the deformation of the solid orthogonal element $\left\{\overrightarrow{PQ}, \overrightarrow{PR}\right\}$, that is computed from Eqs. (3.83) and (3.84). From this figure it becomes clear that the diagonal terms of the relative deformation matrix describe normal strains,

$$
\begin{aligned}
&(PQ') = \sqrt{((1+\varepsilon_{11})dx)^2 + ((\varepsilon_{12}+\omega)dx)^2} \approx \sqrt{dx^2(1+2\varepsilon_{11})} \approx dx(1+\varepsilon_{11}) \\
&\frac{(PQ')-(PQ)}{(PQ)} \approx \frac{(1+\varepsilon_{11})dx - dx}{dx} = \varepsilon_{11}
\end{aligned}
\tag{3.85}
$$

Similarly we get that

$$
\frac{(PR')-(PR)}{(PR)} \approx \varepsilon_{22} \tag{3.86}
$$

Angular strains are given by,

$$
\frac{\pi}{2} - \prec (Q'PR') \approx \frac{(\varepsilon_{12}+\omega)dx}{(1+\varepsilon_{11})dx} + \frac{(\varepsilon_{21}-\omega)dy}{(1+\varepsilon_{22})dy} \approx \varepsilon_{12} + \varepsilon_{21} = 2\varepsilon_{12} = 2\varepsilon_{21} \tag{3.87}
$$

From Fig. 3.4 we see that in case where the strains vanish, $\varepsilon_{ij} = 0$, the deformation of the considered solid orthogonal element is a rigid-body rotation. In Fig. 3.5 we see the relative rotation of the polar material point with respect to the

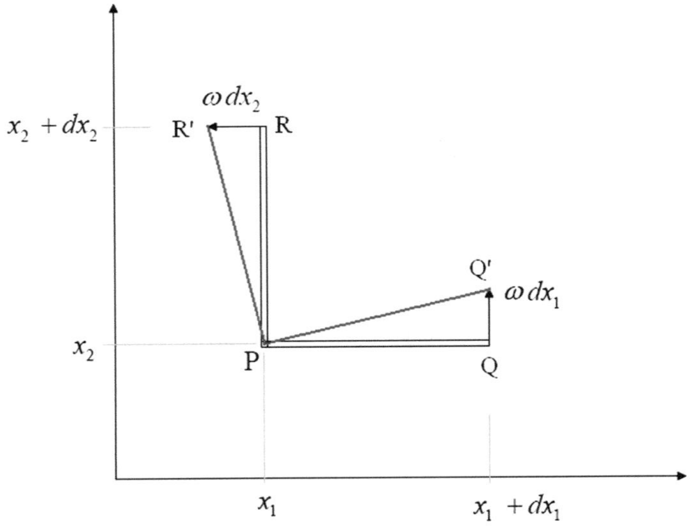

Fig. 3.4 The rotation of a solid orthogonal element

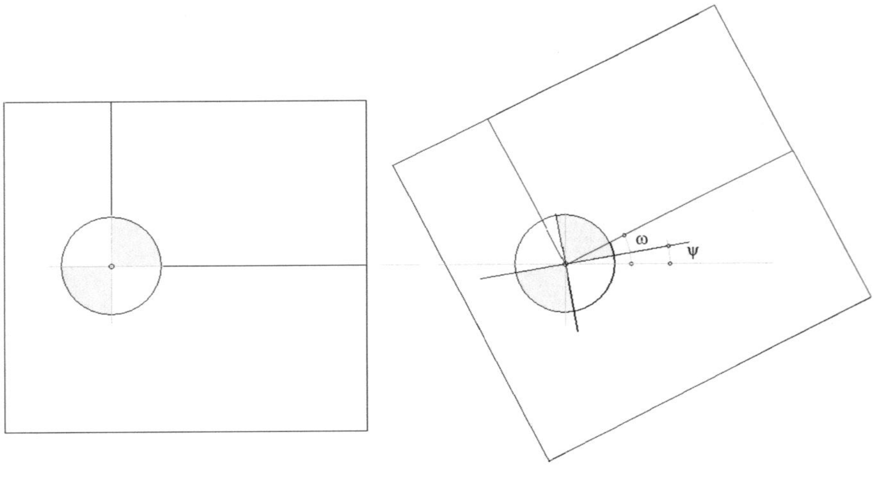

Fig. 3.5 Visualization of the relative spin

rotation of its neighbourhood, caused by the displacement field. In Fig. 3.6, for the visualization of the curvature of the Cosserat deformation we consider the relative rotation of the rigid crosses attached at points Q and R, with respect to the rigid cross attached at point P,

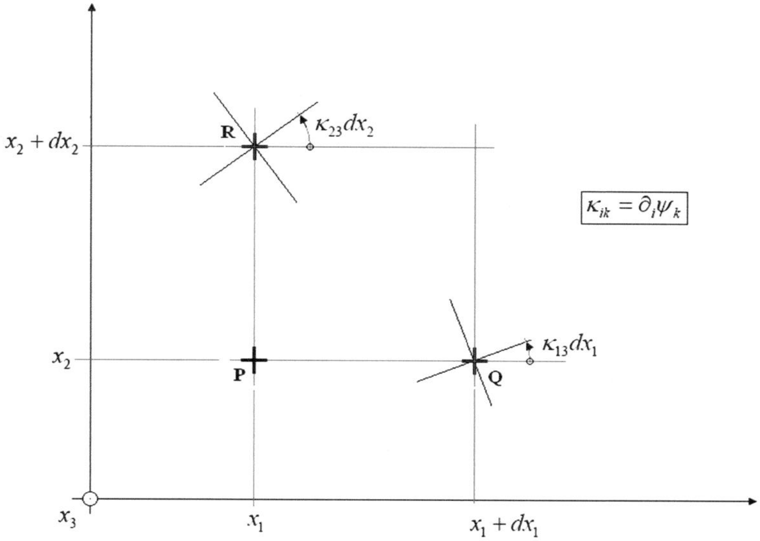

Fig. 3.6 Visualization of the curvature of the deformation

$$\psi(Q) \approx \psi(P) + \kappa_{13}\Delta x_1; \quad \psi(R) \approx \psi(P) + \kappa_{23}\Delta x_2 \tag{3.88}$$

Thus in 2D the curvature tensor is seen as a measure of the bend of the neighbourhood of point P.

3.3 Exercises: Special Orthogonal Curvilinear Coordinates

3.3.1 Polar Cylindrical Coordinates

The polar cylindrical coordinates of a point $P(r, \theta, z)$, are related to its Cartesian coordinates by the following set of equations (Fig. 3.7),

$$\begin{aligned} x &= x^1 = \Theta^1 \cos \Theta^2 = r \cos \theta \quad (0 \leq \theta \leq 2\pi) \\ y &= x^2 = \Theta^1 \sin \Theta^2 = r \sin \theta \\ z &= x^3 = \Theta^3 \end{aligned} \tag{3.89}$$

for $r \in (0, \infty)$ and $\theta \in [0, 2\pi)$.

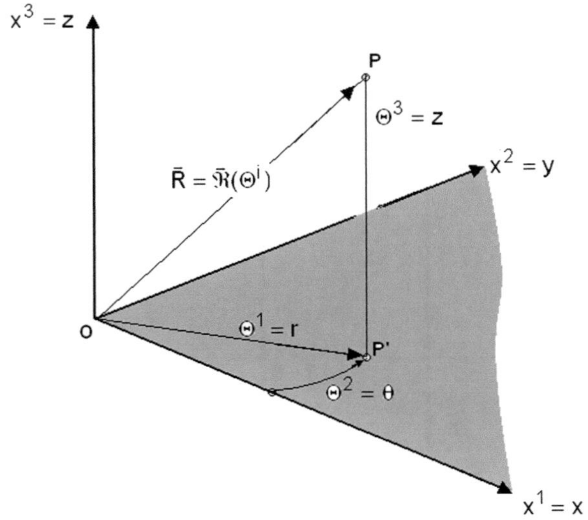

Fig. 3.7 Cartesian and polar cylindrical coordinates

Prove that the deformation tensors in cylindrical polar cylindrical coordinates are as follows,

$$[\nabla u] = \begin{bmatrix} \frac{\partial u_r}{\partial r} & \frac{\partial u_\theta}{\partial r} & \frac{\partial u_z}{\partial r} \\ \frac{1}{r}\frac{\partial u_r}{\partial \theta} - \frac{u_\theta}{r} & \frac{1}{r}\frac{\partial u_\theta}{\partial \theta} + \frac{u_r}{r} & \frac{1}{r}\frac{\partial u_z}{\partial \theta} \\ \frac{\partial u_r}{\partial z} & \frac{\partial u_\theta}{\partial z} & \frac{\partial u_z}{\partial z} \end{bmatrix} \tag{3.90}$$

$$\begin{aligned} sym[\gamma] &= \begin{bmatrix} \varepsilon_{rr} & \varepsilon_{r\theta} & \varepsilon_{rz} \\ \varepsilon_{\theta r} & \varepsilon_{\theta\theta} & \varepsilon_{\theta z} \\ \varepsilon_{zr} & \varepsilon_{z\theta} & \varepsilon_{zz} \end{bmatrix} \\ &= \begin{bmatrix} \frac{\partial u_r}{\partial r} & \frac{1}{2}\left(\frac{1}{r}\frac{\partial u_r}{\partial \theta} + \frac{\partial u_\theta}{\partial r} - \frac{u_\theta}{r}\right) & \frac{1}{2}\left(\frac{\partial u_r}{\partial z} + \frac{\partial u_z}{\partial r}\right) \\ \frac{1}{2}\left(\frac{1}{r}\frac{\partial u_r}{\partial \theta} + \frac{\partial u_\theta}{\partial r} - \frac{u_\theta}{r}\right) & \frac{1}{r}\frac{\partial u_\theta}{\partial \theta} + \frac{u_r}{r} & \frac{1}{2}\left(\frac{\partial u_\theta}{\partial z} + \frac{1}{r}\frac{\partial u_z}{\partial \theta}\right) \\ \frac{1}{2}\left(\frac{\partial u_r}{\partial z} + \frac{\partial u_z}{\partial r}\right) & \frac{1}{2}\left(\frac{\partial u_\theta}{\partial z} + \frac{1}{r}\frac{\partial u_z}{\partial \theta}\right) & \frac{\partial u_z}{\partial z} \end{bmatrix} \end{aligned} \tag{3.91}$$

$$\begin{aligned} &asym[\gamma] \\ &= \begin{bmatrix} 0 & \frac{1}{2}\left(\frac{\partial u_\theta}{\partial r} - \frac{1}{r}\frac{\partial u_r}{\partial \theta} + \frac{u_\theta}{r}\right) - \psi_z & \frac{1}{2}\left(\frac{\partial u_z}{\partial r} - \frac{\partial u_r}{\partial z}\right) + \psi_\theta \\ -\frac{1}{2}\left(\frac{\partial u_\theta}{\partial r} - \frac{1}{r}\frac{\partial u_r}{\partial \theta} + \frac{u_\theta}{r}\right) + \psi_z & 0 & \frac{1}{2}\left(\frac{1}{r}\frac{\partial u_z}{\partial \theta} - \frac{\partial u_\theta}{\partial z}\right) - \psi_r \\ -\frac{1}{2}\left(\frac{\partial u_r}{\partial z} - \frac{\partial u_z}{\partial r}\right) - \psi_\theta & -\frac{1}{2}\left(\frac{1}{r}\frac{\partial u_z}{\partial \theta} - \frac{\partial u_\theta}{\partial z}\right) + \psi_r & 0 \end{bmatrix} \end{aligned} \tag{3.92}$$

$$[\kappa] = \begin{bmatrix} \dfrac{\partial \psi_r}{\partial r} & \dfrac{\partial \psi_\theta}{\partial r} & \dfrac{\partial \psi_z}{\partial r} \\ \dfrac{1}{r}\dfrac{\partial \psi_r}{\partial \theta} - \dfrac{\psi_\theta}{r} & \dfrac{1}{r}\dfrac{\partial \psi_\theta}{\partial \theta} + \dfrac{\psi_r}{r} & \dfrac{1}{r}\dfrac{\partial \psi_z}{\partial \theta} \\ \dfrac{\partial \psi_r}{\partial z} & \dfrac{\partial \psi_\theta}{\partial z} & \dfrac{\partial \psi_z}{\partial z} \end{bmatrix} \tag{3.93}$$

3.3.2 Polar Spherical Coordinates

The polar spherical coordinates of a point be $P(r, \theta, \phi)$ are related to its Cartesian coordinates by the following set of equations (Fig. 3.8)

$$\begin{aligned} x = x^1 &= \Theta^1 \sin \Theta^2 \cos \Theta^3 = r \sin \theta \cos \phi \\ y = x^2 &= \Theta^1 \sin \Theta^2 \sin \Theta^3 = r \sin \theta \sin \phi \\ z = x^3 &= \Theta^1 \cos \Theta^2 = r \cos \theta \end{aligned} \tag{3.94}$$

for $r \in (0, \infty)$, $\theta \in [0, \pi)$ and $\varphi \in [0, 2\pi)$.

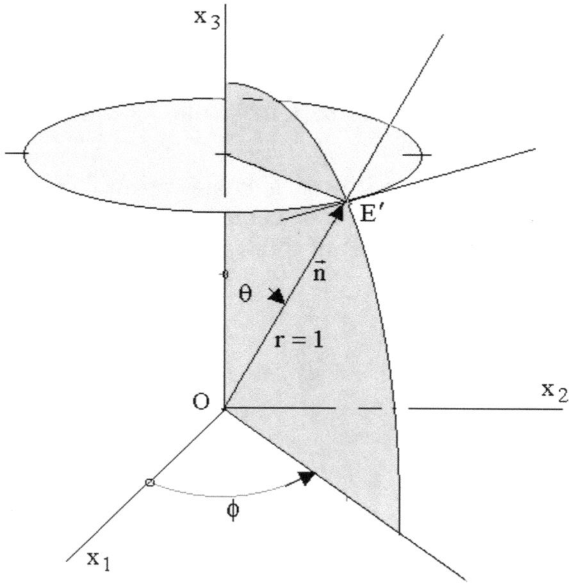

Fig. 3.8 Cartesian and polar spherical coordinates

Prove that the deformation tensors in polar spherical coordinates are as follows,

$$sym[\gamma] = \begin{bmatrix} \frac{\partial u_r}{\partial r} & \frac{1}{2}\left(\frac{1}{r}\frac{\partial u_r}{\partial \theta} + \frac{\partial u_\theta}{\partial r} - \frac{u_\theta}{r}\right) & \frac{1}{2}\left(\frac{\partial u_\phi}{\partial r} + \frac{1}{r\sin\theta}\frac{\partial u_r}{\partial \phi} - \frac{u_\phi}{r}\right) \\ \frac{1}{2}\left(\frac{1}{r}\frac{\partial u_r}{\partial \theta} + \frac{\partial u_\theta}{\partial r} - \frac{u_\theta}{r}\right) & \frac{1}{r}\frac{\partial u_\theta}{\partial \theta} + \frac{u_r}{r} & \frac{1}{2}\left(\frac{1}{r\sin\theta}\frac{\partial u_\theta}{\partial \phi} + \frac{1}{r}\frac{\partial u_\phi}{\partial \theta} - \frac{\cot\theta}{r}u_\phi\right) \\ \frac{1}{2}\left(\frac{\partial u_\phi}{\partial r} + \frac{1}{r\sin\theta}\frac{\partial u_r}{\partial \phi} - \frac{u_\phi}{r}\right) & \frac{1}{2}\left(\frac{1}{r\sin\theta}\frac{\partial u_\theta}{\partial \phi} + \frac{1}{r}\frac{\partial u_\phi}{\partial \theta} - \frac{\cot\theta}{r}u_\phi\right) & \left(\frac{1}{r\sin\theta}\frac{\partial u_\phi}{\partial \phi} + \frac{u_r}{r} + \cot\theta\frac{u_\theta}{r}\right) \end{bmatrix} \tag{3.95}$$

$$asym[\gamma] = \begin{bmatrix} 0 & \frac{1}{2}\left(\frac{\partial u_\theta}{\partial r} - \frac{1}{r}\frac{\partial u_r}{\partial \theta} + \frac{u_\theta}{r}\right) + \psi_\phi & \frac{1}{2}\left(\frac{\partial u_\phi}{\partial r} - \frac{1}{r\sin\theta}\frac{\partial u_r}{\partial \phi} + \frac{u_\phi}{r}\right) - \psi_\theta \\ -\frac{1}{2}\left(\frac{\partial u_\theta}{\partial r} - \frac{1}{r}\frac{\partial u_r}{\partial \theta} + \frac{u_\theta}{r}\right) - \psi_\phi & 0 & \frac{1}{2}\left(\frac{1}{r}\frac{\partial u_\phi}{\partial \theta} - \frac{1}{r\sin\theta}\frac{\partial u_\theta}{\partial \phi} + \frac{\cot\theta}{r}u_\phi\right) + \psi_r \\ -\frac{1}{2}\left(\frac{\partial u_\phi}{\partial r} - \frac{1}{r\sin\theta}\frac{\partial u_r}{\partial \phi} + \frac{u_\phi}{r}\right) + \psi_\theta & -\frac{1}{2}\left(\frac{1}{r}\frac{\partial u_\phi}{\partial \theta} - \frac{1}{r\sin\theta}\frac{\partial u_\theta}{\partial \phi} + \frac{\cot\theta}{r}u_\phi\right) - \psi_r & 0 \end{bmatrix} \tag{3.96}$$

$$[\kappa] = \begin{bmatrix} \frac{\partial \psi_r}{\partial r} & \frac{\partial \psi_\theta}{\partial r} & \frac{\partial \psi_\phi}{\partial r} \\ \frac{1}{r}\frac{\partial \psi_r}{\partial \theta} - \frac{\psi_\theta}{r} & \frac{1}{r}\frac{\partial \psi_\theta}{\partial \theta} + \frac{\psi_r}{r} & \frac{1}{r}\frac{\partial \psi_\phi}{\partial \theta} \\ \frac{1}{r\sin\theta}\frac{\partial \psi_r}{\partial \phi} - \frac{\psi_\phi}{r} & \frac{1}{r\sin\theta}\frac{\partial \psi_\theta}{\partial \phi} - \frac{1}{r\tan\theta}\psi_\theta & \frac{1}{r\sin\theta}\frac{\partial \psi_\phi}{\partial \phi} + \frac{\psi_r}{r} + \frac{1}{r\tan\theta}\psi_\theta \end{bmatrix} \tag{3.97}$$

3.4 Integrability Conditions and Compatibility Equations

3.4.1 Formulation in General Curvilinear Coordinates

Let (Γ) be a curve in space that is passing through points P_0 and P. Starting from point P_0 we can compute the value of one of the Cosserat particle kinematic fields, say the particle rotation, by means of a line integral that is evaluated along the considered curve (Γ). Thus from

$$\boldsymbol{\psi}(P) = \boldsymbol{\psi}(P_0) + \int_{P_0}^{P} \boldsymbol{\psi}_{,k} d\Theta^k \tag{3.98}$$

and Eq. (3.20) we get

$$\begin{aligned} \boldsymbol{\psi}(P_0) &= \boldsymbol{\psi}_0 \\ \boldsymbol{\psi}(P) &= \boldsymbol{\psi}_0 + \int_{P_0}^{P} \boldsymbol{\kappa}_k d\Theta^k \end{aligned} \tag{3.99}$$

For uniqueness purposes we may require that the value of the Cosserat rotation at point P, as computed from Eq. (3.99), is independent of the particular choice of the curve (Γ) that joins the points P_0 and P; assuming that at point P_0 the value of $\boldsymbol{\psi}$ is known. According to the fundamental theorem of Tensor Analysis, the sufficient and necessary condition for this integrability requirement is that

$$rot\boldsymbol{\kappa}_k = \mathbf{0} \tag{3.100}$$

or in components,

$$\overset{(1)}{\boldsymbol{J}^i} = e^{ijk}\boldsymbol{\kappa}_{k,j} = 0 \tag{3.101}$$

The first-order system $\overset{(1)}{\boldsymbol{J}^i}$ is called the 1st incompatibility form [4].
With

$$\boldsymbol{\kappa}_{k,j} = \kappa_{k|j}^{\cdot l}\boldsymbol{g}_l \tag{3.102}$$

we get

$$\overset{(1)}{\boldsymbol{J}^i} = e^{ijk}\boldsymbol{\kappa}_{k|j}^{\cdot l}\boldsymbol{g}_l = \overset{(1)}{I^{il}}\,\boldsymbol{g}_l \tag{3.103}$$

Thus, Eq. (3.101) yields the 1st set of compatibility equations [4],

$$\overset{(1)}{I^{il}} = e^{ijk}\kappa_{k|j}^{\cdot l} = 0 \tag{3.104}$$

Similarly from,

$$\boldsymbol{u}(P) = \boldsymbol{u}(P_0) + \int_{P_0}^{P} \boldsymbol{u}_{,k} d\Theta^k = \boldsymbol{u}_0 + \int_{P_0}^{P} (\gamma_k - \boldsymbol{g}_k \times \boldsymbol{\psi}) d\Theta^k \tag{3.105}$$

and Eq. (3.4) we get,

$$\boldsymbol{u}(P) = \boldsymbol{u}_0 + \boldsymbol{\psi}_0 \times (\boldsymbol{R} - \boldsymbol{R}_0) + \int_{\boldsymbol{P}_0}^{\boldsymbol{P}_1} (\gamma_k + (\boldsymbol{R} - \boldsymbol{R}_0) \times \boldsymbol{\kappa}_k) d\Theta^k \tag{3.106}$$

The integrability of Eq. (3.106) results to the following condition

$$e^{ijk}(\gamma_k + (\boldsymbol{R} - \boldsymbol{R}_0) \times \boldsymbol{\kappa}_k)_{,j} = 0 \tag{3.107}$$

or

$$e^{ijk}\left(\gamma_{k,j} + \boldsymbol{R}_{,j} \times \boldsymbol{\kappa}_k + (\boldsymbol{R} - \boldsymbol{R}_0) \times \boldsymbol{\kappa}_{k,j}\right) = 0 \tag{3.108}$$

Due to Eqs. (3.101) and (3.4), Eq. (3.108) yields the following condition,

$$\overset{(2)}{\boldsymbol{J}^i} = e^{ijk}\left(\gamma_{k,j} + \boldsymbol{g}_j \times \boldsymbol{\kappa}_k\right) = \boldsymbol{0} \tag{3.109}$$

The first-order system $\overset{(2)}{\boldsymbol{J}^i}$ is called the 2nd incompatibility form. With

$$\gamma_{k,j} = \gamma_{klj}\boldsymbol{g}^l \tag{3.110}$$

and

$$\boldsymbol{g_j} \times \boldsymbol{\kappa_k} = \boldsymbol{g_j} \times \kappa_k^{\cdot m}\boldsymbol{g}_{\mathrm{m}} = \boldsymbol{e}_{ljm}\kappa_k^{\cdot m}\boldsymbol{g}^l \tag{3.111}$$

we get

$$\overset{(2)}{\boldsymbol{J}^i} = e^{ijk}\left(\gamma_{kl|j} + e_{ljm}\kappa_k^{\cdot m}\right)\boldsymbol{g}^l = \overset{(2)}{I^i_{\cdot l}}\,\boldsymbol{g}^l \tag{3.112}$$

Thus Eq. (3.109) yields [4],

$$\overset{(2)}{I^i_{\cdot l}} = e^{ijk}\left(\gamma_{kl|j} + e_{ljm}\kappa_k^{\cdot m}\right) = 0 \tag{3.113}$$

The motion is called to be an incompatible one, if the "incompatibilities" $\overset{(1)}{I^{il}}$ and $\overset{(2)}{I^i_{\cdot l}}$ are not zero.

In view of the above derivations and following Kessel [2], we define the rotor of the generalized-displacement gradient $\bar{\bar{H}}$,

$$\bar{\bar{H}} = \begin{pmatrix} \kappa_i^{\cdot k}\boldsymbol{g}^i \otimes \boldsymbol{g}_k \\ \gamma_{ik}\boldsymbol{g}^i \otimes \boldsymbol{g}^k \end{pmatrix} \quad \Rightarrow \quad Rot\bar{\bar{H}} := -e^{ikj}\begin{pmatrix} \kappa_{k|j}^{\cdot l}\boldsymbol{g}_i \otimes \boldsymbol{g}_l \\ \left(\gamma_{kl|j} + \kappa_k^{\cdot m}e_{jml}\right)\boldsymbol{g}_i \otimes \boldsymbol{g}^l \end{pmatrix} \tag{3.114}$$

and with that the compatibility Eqs. (3.104) and (3.113) become,

$$Rot\bar{\bar{H}} = \bar{\bar{0}} \tag{3.115}$$

Indeed, with Eq. (3.36) we get formally that these newly defined differential operators on the generalized-displacement gradient satisfy the well-known identity,

$$\bar{\bar{H}} = Grad\bar{K} \quad \Rightarrow \quad RotGrad\bar{K} = \bar{\bar{0}} \tag{3.116}$$

3.4.2 Compatibility Equations in Cartesian Coordinates

In Cartesian coordinates the compatibility conditions, Eqs. (3.104) and (3.113) become,

$$\overset{(1)}{I}_{kl} = \varepsilon_{kpq}\partial_p\kappa_{ql} = 0 \tag{3.117}$$

$$\overset{(2)}{I}_{pi}\ \varepsilon_{pjk}\partial_j\gamma_{ki} + \delta_{pi}\kappa_{ll} - \kappa_{ip} = 0 \tag{3.118}$$

Explicitly these compatibility equations read as follows,

$$\begin{aligned}
&\overset{(1)}{I}_{11} = 0 = \varepsilon_{1pq}\partial_p\kappa_{q1} = \varepsilon_{123}\partial_2\kappa_{31} + \varepsilon_{132}\partial_3\kappa_{21} = \partial_2\kappa_{31} - \partial_3\kappa_{21}\\
&\overset{(1)}{I}_{12} = 0 = \varepsilon_{1pq}\partial_p\kappa_{q2} = \varepsilon_{123}\partial_2\kappa_{32} + \varepsilon_{132}\partial_3\kappa_{22} = \partial_2\kappa_{32} - \partial_3\kappa_{22}\\
&\ldots
\end{aligned} \tag{3.119}$$

and

$$\begin{aligned}
&\overset{(2)}{I}_{11} = 0 = \varepsilon_{1jk}\partial_j\gamma_{k1} - \kappa_{11} + \delta_{11}\kappa_{kk} = \partial_2\gamma_{31} - \partial_3\gamma_{21} + \kappa_{22} + \kappa_{33}\\
&\overset{(2)}{I}_{12} = 0 = \varepsilon_{1jk}\partial_j\gamma_{k2} - \kappa_{21} + \delta_{21}\kappa_{kk} = \partial_2\gamma_{32} - \varepsilon_{132}\partial_3\gamma_{22} - \kappa_{21}\\
&\ldots
\end{aligned} \tag{3.120}$$

If κ_{ij} are the components of the gradient of a vector field ψ_k, then the compatibility Eqs. (3.119) reduce to the differentiability conditions for the named vector field,

$$\begin{aligned}
&\overset{(1)}{I}_{11} = 0 = \partial_2\partial_3\psi_1 - \partial_3\partial_2\psi_1\\
&\overset{(1)}{I}_{12} = 0 = \partial_2\partial_3\psi_2 - \partial_3\partial_2\psi_2\\
&\ldots
\end{aligned} \tag{3.121}$$

Similarly, if the γ_{ij} are given by Eqs. (3.59), then the compatibility Eqs. (3.120) reduce to the differentiability conditions for the vector field u_i,

$$\begin{aligned}
\overset{(2)}{I}_{11} &= 0 = \partial_2(\partial_3 u_1 - \varepsilon_{31l}\psi_l) - \partial_3(\partial_2 u_1 - \varepsilon_{21l}\psi_l) + \partial_2\psi_2 + \partial_3\psi_3\\
&= \partial_2\partial_3 u_1 - \partial_3\partial_2 u_1 - \partial_2\psi_2 - \partial_3\psi_3 + \partial_2\psi_2 + \partial_3\psi_3 = \partial_2\partial_3 u_1 - \partial_3\partial_2 u_1\\
&\ldots
\end{aligned} \tag{3.122}$$

Finally in a 2D setting, the above compatibility conditions yield

$$\overset{(1)}{I}_{33} = \varepsilon_{3pq}\partial_p\kappa_{q3} = \varepsilon_{321}\partial_2\kappa_{13} + \varepsilon_{312}\partial_1\kappa_{23} \\ = -\partial_2\kappa_{13} + \partial_1\kappa_{23} = 0 \tag{3.123}$$

$$\begin{aligned} \overset{(2)}{I}_{31} &= \varepsilon_{3jk}\partial_j\gamma_{k1} - \kappa_{13} = \varepsilon_{312}\partial_1\gamma_{21} + \varepsilon_{321}\partial_2\gamma_{11} - \kappa_{13} \\ &= \partial_1\gamma_{21} - \partial_2\gamma_{11} - \kappa_{13} = 0 \\ \overset{(2)}{I}_{32} &= \varepsilon_{3jk}\partial_j\gamma_{k2} - \kappa_{23} = \varepsilon_{312}\partial_1\gamma_{22} + \varepsilon_{321}\partial_2\gamma_{12} - \kappa_{23} \\ &= \partial_1\gamma_{22} - \partial_2\gamma_{12} - \kappa_{23} = 0 \end{aligned} \tag{3.124}$$

3.5 Kinematical Compatibility Conditions for Stationary Discontinuities

The mathematical treatment of discontinuity surfaces as applied to continuum mechanics can be found in the textbooks of Thomas [5] and Vardoulakis and Sulem [6].

Let $[\cdot]$ denote the jump of a quantity across a discontinuity surface S_D,

$$[Z] = Z^+ - Z^- \tag{3.125}$$

where Z^+ and Z^- are the one-sited limits of the (scalar, vector or tensor) function Z on S_D, whose positive side is determined by the unit outward normal vector $\boldsymbol{n}$.

In the considered context we will be interested in the formulation of compatibility conditions across material discontinuity surfaces, i.e. discontinuity surfaces that move attached always to the same material particles. In that sense we have to introduce the time dependence in the argument list of the kinematic fields that describe the motion of the particle by setting

$$\begin{aligned} \psi^i &= \psi^i\left(\Theta^k, t\right) \\ u_i &= u_i\left(\Theta^k, t\right) \end{aligned} \tag{3.126}$$

In Eulerian description the particle is moved with the velocity,

$$v_i = \frac{\partial u_i}{\partial t} + u_{i|k}v^k \quad \Rightarrow \quad \left(g_{ik} - u_{i|k}\right)v^k = \frac{\partial u_i}{\partial t} \tag{3.127}$$

We assume that across S_D the particle displacement and its 1st order derivatives are continuous,

$$[u_i] = 0; \quad \left[\frac{\partial u_i}{\partial t}\right] = 0; \quad [u_{i,k}] = 0 \quad \Rightarrow \quad [u_{i|k}] = 0 \tag{3.128}$$

From Eqs. (3.127) and (3.128) we get that the velocity vector is continuous,

$$[v_i] = 0 \tag{3.129}$$

This implies the following Maxwell conditions,

$$[v_{k|i}] = [v_{k,i}] = b_k n_i \tag{3.130}$$

With the velocity being continuous, the considered material discontinuity surface S_D moves with the normal velocity of the particles that are attached to it,

$$c = v_n = v_i n^i \tag{3.131}$$

If we assume that the particle rotation is also continuous across S_D,

$$[\psi^i] = 0 \tag{3.132}$$

then the corresponding kinematical compatibility conditions for the 1st order partial derivatives of ψ^i are [5],

$$\left[\frac{\partial \psi^i}{\partial t}\right] = -\lambda^i c; \quad [\psi^i_{|k}] = \lambda^i n_k \tag{3.133}$$

or due to Eq. (3.131),

$$\left[\frac{\partial \psi^i}{\partial t}\right] = -\lambda^i v_n; \quad [\psi^i_{|k}] = \lambda^i n_k \tag{3.134}$$

Let the rate of particle rotation be denoted as

$$w^i = \frac{\partial \psi^i}{\partial t} + v^k \psi^i_{|k} \tag{3.135}$$

From the compatibility conditions (3.134) and Eq. (3.135) follows that the jump in particle spin must vanish [7],

$$[w^i] = -\lambda^i v_n + \lambda^i n_k v^k = 0 \tag{3.136}$$

This implies in turn the following Maxwell conditions,

$$\left[w^{k}_{\cdot\,|i}\right] = a^k n_i \tag{3.137}$$

Within the frame of a small strain theory, material time differentiation of the deformation measures, Eqs. (3.32) and (3.33), leads to the definitions of the corresponding rate-of-deformation measures denoted as: (a) The distortion-rate tensor

$$K_i^{\cdot k} = w^{k}_{\cdot\,|i} \tag{3.138}$$

and (b) the rate of deformation tensor,

$$\Gamma_{ik} = v_{k|i} - e_{ikl} w^l \tag{3.139}$$

From the continuity requirements, Eqs. (3.129), (3.130) and (3.136), (3.137) we get the compatibility conditions for the rate of deformation measures,

$$\left[K_i^{\cdot k}\right] = a^k n_i \tag{3.140}$$

and

$$[\Gamma_{ik}] = b_k n_i \tag{3.141}$$

In certain cases we may be forced to consider the existence of strong discontinuities of the particle rotation vector, that are identified as *strong particle-rolling* discontinuities. This means that we may have to assume that,

$$\left[\psi^i\right] = r^i \neq 0 \tag{3.142}$$

In this case the corresponding geometrical compatibility conditions for the distortions are rather involved expressions that account also for the curvature of the discontinuity surface. The derivation of such compatibility equations for strongly discontinuous fields can be found in [5] and will be omitted here.

References

1. Schaefer, H. (1967). Analysis der Motorfelder in Cosserat-Kontinuum. *Zeitschrift für Angewandte Mathematik und Mechanik, 47,* 319–332.
2. Kessel, S. (1967). Stress functions and loading singularities for the infinitely extended linear elastic-isotropic Cosserat Continuum. In *Proceedings of the IUTAM-symposium on the generalized Cosserat continuum and the continuum theory of dislocations with applications.* Stuttgart: Springer.

3. Schaefer, H. (1962). Versuch einer Elastizitätstheorie des zweidimensionalen ebenen Cosserat-Kontinuums. In: *Miszellaneen der angewandten Mechanik* (pp. 277–292). Akademie-Verlag: Berlin.
4. Günther, W. (1958). Zur Statik und Kinematik des Cosseratschen Kontinuums. *Abhandlungen der Braunschweigische Wissenschaftliche Gesellschaft, 10,* 195–213.
5. Thomas, T. Y. (1961). *Plastic flow and fracture* (Vol. 2). Academic Press.
6. Vardoulakis, I., & Sulem, J. (1995). *Bifurcation analysis in geomechanics*. Blackie Academic & Professional.
7. Dietsche, A., & Willam, K. (1997). Boundary effects in elasto-plastic Cosserat continua. *International Journal of Solids and Structures, 34*(7), 877–893.

Chapter 4
Cosserat Continuum Statics

Abstract In this chapter conservation considerations are firstly introduced. The virtual work principle, together with equilibrium equations in generalized curvilinear coordinate systems, are briefed under the auspices of the new mathematical representation, with working examples of how they reduce in cartesian, polar cylindrical and spherical coordinate systems. Finally, the definition of the concept of the traction motor—in accordance to the concept of the traction vector through Cauchy's tetrahedron—is detailed.

It is well known [1] that the statics of rigid bodies can be developed axiomatically and independently of dynamics. Since rigid bodies interact by pairs of opposed forces and couples, we keep from Newton's laws only *lex tertia*. Along this line of thought we introduce here the notions of stress and couple stress in Cosserat continua by resorting to the principle of virtual work.

4.1 The Virtual Work Equation

We consider a Cosserat continuum **B**, that occupies a domain with volume V that has the boundary ∂V. Body **B** is assumed to be in a state of stress in static equilibrium. In order to formulate the equilibrium conditions we consider fields $\delta\psi^i(\Theta^k)$ and $\delta u_i(\Theta^k)$, that are defined uniquely at all points of the given body. These fields will be called virtual particle rotation and virtual particle displacement fields, respectively, and it will be assumed that they are sufficiently differentiable.

We define the virtual curvature and relative deformation tensors,

$$\delta\kappa_i^{\cdot k} = \delta\psi^k_{\cdot|i} \tag{4.1}$$

$$\delta\gamma_{ik} = \delta u_{k|i} + e_{kil}\delta\psi^l \tag{4.2}$$

I. Vardoulakis, *Cosserat Continuum Mechanics*, Lecture Notes in Applied and Computational Mechanics 87, https://doi.org/10.1007/978-3-319-95156-0_4

We define fields $\sigma^{ij}(\Theta^k)$ and $\mu^i_{\cdot j}(\Theta^k)$, through the so called virtual work of the internal forces, that is in turn defined per unit volume of the considered Cosserat continuum,

$$\delta w^{(\mathrm{int})} = \sigma^{ik}\delta\gamma_{ik} + \mu^i_{\cdot k}\delta\kappa_i^{\cdot k} \tag{4.3}$$

We assume that $\delta w^{(\mathrm{int})}$ is an invariant scalar quantity. The tensor $\delta\gamma_{ij}$ is a covariant 2nd order tensor and the tensor $\delta\kappa_i^{\cdot k}$ is a mixed co-contravariant 2nd order tensor. For $\delta w^{(\mathrm{int})}$ to be invariant, σ^{ij} must be a full contravariant 2nd order tensor and $\mu^i_{\cdot j}$ a mixed, conte-covariant 2nd order tensor. Thus the quantities σ and μ in Eq. (4.3) are tensors and are called the stress- and couple-stress tensors, respectively.

Remarks At this point we should remark that in some occasions in the literature alternative but equivalent definitions for the virtual work of internal forces can be found. For example one can start from the following definition [2],

$$\delta w^{(\mathrm{int})} = \sigma^{ik}\delta\tilde{\gamma}_{ki} + \mu^i_{\cdot k}\delta\tilde{\kappa}^k_{\cdot i} \tag{4.4}$$

This definition is based on deformation measures that are simply the transposed of the ones we have have introduced here,

$$\delta\tilde{\kappa}^k_{\cdot i} = \delta\psi^{\cdot k}_{|i}; \quad \delta\tilde{\gamma}_{ik} = \delta u_{i|k} + e_{ikl}\delta\psi^l \tag{4.5}$$

Alernatively one may define the virtual work of internal forces as follows [3, 4],

$$\delta w^{(\mathrm{int})} = \tilde{\sigma}^{ik}\delta\tilde{\gamma}_{ik} + \tilde{\mu}^i_{\cdot k}\delta\tilde{\kappa}^i_{\cdot k} \tag{4.6}$$

This definition is based on stress and deformation measures that are simply the transposed of the ones we have introduced here.

We recall also that an *intensive* property (also called a bulk property) is a physical property of a system that does not depend on the system's size or the amount of material (mass) in the system. By contrast, an *extensive* property of a system does depend on the system size or the amount of material in the system. From the point of view of continuum thermodynamics the stress-and couple stress tensors are intensive quantities, that are dual in energy to the relative deformation tensor and distortion tensor, respectively, that are in turn the corresponding mechanical intensive quantities of the considered continuum.

We decompose additively the virtual relative deformation into symmetric and antisymmetric part,

$$\delta\gamma_{ij} = \delta\gamma_{(ij)} + \delta\gamma_{[ij]} \tag{4.7}$$

where

$$\begin{aligned}\delta\gamma_{(ij)} &= \frac{1}{2}\left(\delta\gamma_{ij} + \delta\gamma_{ji}\right)\\ \delta\gamma_{[ij]} &= \frac{1}{2}\left(\delta\gamma_{ij} - \delta\gamma_{ji}\right)\end{aligned} \tag{4.8}$$

Let

$$\delta\varepsilon_{ij} = \frac{1}{2}\left(\delta u_{j|i} + \delta u_{i|j}\right) \tag{4.9}$$

$$\delta\omega_{ij} = \frac{1}{2}\left(\delta u_{i|j} - \delta u_{j|i}\right) = -e_{ijk}\delta\omega^k \tag{4.10}$$

where $\delta\omega^k$ is the axial vector that corresponds to the non-symmetric part of the virtual displacement gradient. Thus,

$$\delta\gamma_{ij} = \delta\varepsilon_{ij} + e_{ijk}\left(\delta\omega^k - \delta\psi^k\right) \tag{4.11}$$

Similarly, we decompose additively the stress tensor into symmetric and antisymmetric part,

$$\sigma^{ij} = \sigma^{(ij)} + \sigma^{[ij]} \tag{4.12}$$

where

$$\begin{aligned}\sigma^{(ij)} &= \frac{1}{2}\left(\sigma^{ij} + \sigma^{ji}\right)\\ \sigma^{[ij]} &= \frac{1}{2}\left(\sigma^{ij} - \sigma^{ji}\right)\end{aligned} \tag{4.13}$$

With this decomposition the virtual work of the internal forces, Eq. (4.3), becomes,

$$\begin{aligned}\delta w^{(\text{int})} &= \sigma^{(ij)}\delta\varepsilon_{ij} + \sigma^{[ij]}\delta\gamma_{[ij]} + \mu^{i}_{\cdot k}\delta\kappa_i^{\cdot k}\\ &= \sigma^{(ij)}\delta\varepsilon_{ij} + \sigma^{[ij]}e_{ijk}\left(\delta\omega^k - \delta\psi^k\right) + \mu^{i}_{\cdot k}\delta\kappa_i^{\cdot k}\end{aligned} \tag{4.14}$$

The antisymmetric part of the stress tensor is dual in work to the relative spin,

$$\sigma^{[ij]}\delta\gamma_{[ij]} = 2t_k^*\left(\delta\omega^k - \delta\psi^k\right) \tag{4.15}$$

where t_i^* is the axial vector that corresponds to the non-symmetric part of the stress tensor.

$$t_i^* = \frac{1}{2}e_{ijk}\sigma^{jk} = \frac{1}{2}e_{ijk}\sigma^{[jk]} \Leftrightarrow \sigma^{[jk]} = e^{ijk}t_i^* \tag{4.16}$$

With this remark Eq. (4.14) becomes,

$$\delta w^{(\mathrm{int})} = \sigma^{(ij)}\delta\varepsilon_{ij} + 2t_i^*\left(\delta\omega^i - \delta\psi^i\right) + \mu^i_{\cdot k}\delta\kappa_i^{\cdot k} \tag{4.17}$$

This equation demonstrates that in a Cosserat continuum the virtual work of the internal forces is defined in such a way that: (a) The symmetric part of the stress tensor is dual to the strain. (b) The antisymmetric part of the stress tensor is dual to the relative particle spin. For this reason we may call $\sigma^{[ij]}$ the "relative" stress tensor. (c) The couple stress tensor is dual to the distortion tensor.

The *work of the internal forces* is defined as the integral of the work density function $\delta w^{(\mathrm{int})}$ over the volume V,

$$\delta W^{(\mathrm{int})} = \int_{(V)} \delta w^{(\mathrm{int})} dV \tag{4.18}$$

For the formulation of the principle of virtual work, we must define also the virtual work of the "external" actions. In particular we assume that on the Cosserat continuum body three types of external actions are applied: (a) Volume forces $f^i dV$, where dV is the volume element. (b) Surface tractions $t^i dS$. (c) Surface couples $m_i dS$. In these expressions dS is the surface element. In general one may assume the existence of body couples as well; this case will be disregarded here.

The bounding material surface ∂V of a material volume V is seen as a two dimensional, piecewise smooth particle manifold, with each particle of that manifold possessing two vectorial degrees of freedom, the one of particle rotation and that of particle displacement.

For clarity we use here for the description of the boundary conditions the natural curvilinear coordinates of the surface. The position of any point $P \in \partial V$ is given by its surface coordinates α^1 and α^2. The position of points in space inside and outside that surface are described by their normal distance from it, that is given by the coordinate α^3; i.e. the coordinate that is measured positive along the outward normal to the surface. At the arbitrary point $P(\alpha^1, \alpha^2, 0)$ on the surface we can define the corresponding covariant basis, $(\alpha_1, \alpha_2, \alpha_3)$, as is shown in Fig. 4.1. From that basis we construct the corresponding contravariant basis $(\alpha^1, \alpha^2, \alpha^3)$. With this notation we can express admissible sets of boundary conditions by assigning continually a number of individual components of the kinematic and static vector properties of the surface particles, the components themselves being defined with respect to aforementioned covariant and contravariant surface vector bases. For example, a set of admissible boundary conditions at point $P(\alpha^1, \alpha^2, 0)$ could be the following,

$$\left\{P \in \partial V : \left\{S_D : (p)_P = \begin{pmatrix} \times & \times & \psi^3 \\ \times & \times & u_3 \end{pmatrix}_P\right\} \cup \left\{S_N : (q)_P = \begin{pmatrix} m_1 & m_2 & \times \\ t^1 & t^2 & \times \end{pmatrix}_P\right\}\right\} \tag{4.19}$$

Fig. 4.1 Local coordinates in a point at the bounding surface

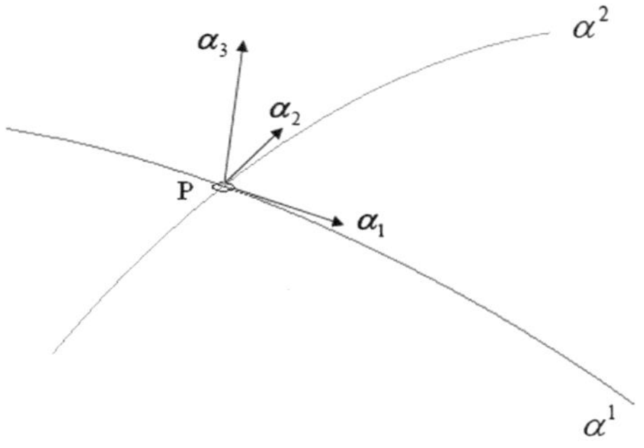

The sets S_D and S_N are called here the Dirichlet and the Neumann set, respectively. In the example given above by Eq. (4.19) in the neighbourhood of point $P(\alpha^1, \alpha^2, 0)$ and along the α^α-surface lines ($\alpha = 1, 2$) at least locally tractions and couples are prescribed, whereas normal to the surface the particle displacement and spin are restricted. This example illustrates also the assumption, that if some information, say $\psi^3 = p_{31}$, is given for a surface particle at $P(\alpha^1, \alpha^2, 0)$, no information concerning $m_3 = q_{31}$ can be given at this particle et vice versa. This is exactly the mathematical meaning of the "empty slot" symbols $\times$, used in the corresponding matrices of Eq. (4.19). The corresponding entries to these empty slots are called reactive constraints.

On the basis of the above definitions we define a functional that is called the *virtual work of external forces*,

$$\delta W^{(ext)} = \int_{(V)} f^i \delta u_i dV + \int_{(S_N)} \left(m_i \delta \psi^i + t^i \delta u_i\right) dS \tag{4.20}$$

With these definitions we remark that the second integral on the r.h.s. of Eq. (4.20) is a generalized integral of the Lebesgue type. In order to remove this difficulty, we assume that the virtual kinematics vanish on the complementary part of the boundary; i.e. we assume that,

$$on\ S_D : \ \delta\psi^i = 0 \wedge \delta u_i = 0 \tag{4.21}$$

We assume that these data are continuously extended into V and on the disjoint parts of the boundary. Thus, whatever the values of the reactive constraints are on S_D, the functional, Eq. (4.20), can be continuously extended over the whole boundary, and

$$\delta W^{(ext)} = \int_{(V)} f^i \delta u_i dV + \int_{(\partial V)} \left(m_i \delta \psi^i + t^i \delta u_i\right) dS \tag{4.22}$$

where the second integral on the r.h.s. of Eq. (4.22) is a normal Riemann surface integral.

On the basis of the above definitions, the principle of virtual work in a Cosserat continuum is defined as follows:

Definition *The system* $\left\{\sigma^{ij}, \mu^{i}_{\cdot j}; f^{i}, t^{i}, m_{i}\right\}$ *is called an equilibrium set, if, for any choice of the virtual fields of particle displacement and rotation that satisfy* Eq. (4.21), *the virtual work equation holds,*

$$\delta W^{(ext)} = \delta W^{(\mathrm{int})} \tag{4.23}$$

From Eq. (4.23) and the definitions for the virtual work of internal- and external forces, Eqs. (4.17), (4.3) and (4.22), we obtain the following integral equation,

$$\int_{(V)} f^{i}\delta u_{i}dV + \int_{(\partial V)} t^{i}\delta u_{i}dS + \int_{(\partial V)} m_{i}\delta\psi^{i}dS = \int_{(V)} \left(\sigma^{ik}\delta\gamma_{ik} + \mu^{i}_{\cdot k}\delta\kappa_{i}^{\cdot k}\right)dV \tag{4.24}$$

This is the virtual work equation for a Cosserat continuum in the absence of body couples.

4.2 Equilibrium Equations

4.2.1 General Curvilinear Coordinates

We remark first that the density of the virtual work of the internal forces can be written as follows,

$$\begin{aligned}\delta w^{(\mathrm{int})} &= \sigma^{ik}\left(\delta u_{k|i} - e_{ikl}\delta\psi^{l}\right) + \mu^{i}_{\cdot k}\delta\psi^{k}_{\cdot\ |i} \\ &= \left(\sigma^{ik}\delta u_{k} + \mu^{i}_{\cdot k}\delta\psi^{k}\right)_{|i} - \left(\sigma^{ik}_{|i}\delta u_{k} + \mu^{i}_{\cdot k|i}\delta\psi^{k}\right) - e_{ikl}\sigma^{ik}\delta\psi^{l}\end{aligned} \tag{4.25}$$

With the notation,

$$q^{i} = \sigma^{ik}\delta u_{k} + \mu^{i}_{\cdot k}\delta\psi^{k} \tag{4.26}$$

and with the use of Gauss' theorem we get,

$$\int_{V} q^{i}_{\cdot|i}\, dV = \int_{\partial V} q^{i}n_{i}\, dS \tag{4.27}$$

and,

$$\int_V \left(\sigma^{ik}\delta u_k + \mu^{i}_{.k}\delta\psi^k\right)_{|i} dV = \int_{\partial V} \left(\sigma^{ik}\delta u_k + \mu^{i}_{.k}\delta\psi^k\right) n_i dS \tag{4.28}$$

With Eq. (4.28) the virtual work Eq. (4.24) becomes

$$\begin{aligned}\int_{(V)} f^k\delta u_k dV + \int_{(\partial V)} t^k\delta u_k dS + \int_{(\partial V)} m_k\delta\psi^k dS = \int_{\partial V} \left(\sigma^{ik}\delta u_k + \mu^{i}_{.k}\delta\psi^k\right) n_i dS \\ - \int_{(V)} \sigma^{ik}_{|i}\delta u_k dV - \int_{(V)} \left(\mu^{i}_{.l|i} + e_{ikl}\sigma^{ik}\right)\delta\psi^l dV\end{aligned} \tag{4.29}$$

or

$$\begin{aligned}\int_{(V)} \left(\sigma^{ik}_{|i} + f^k\right)\delta u_k dV + \int_{(V)} \left(\mu^{i}_{.k|i} + e_{ilk}\sigma^{il}\right)\delta\psi^k dV \\ = \int_{(\partial V)} \left(\sigma^{ik}n_i - t^k\right)\delta u_k dS + \int_{(\partial V)} \left(\mu^{i}_{.k}n_i - m_k\right)\delta\psi^k dS\end{aligned} \tag{4.30}$$

The test functions $\delta u_k(\Theta^i)$ and $\delta\psi^k(\Theta^i)$ can be chosen arbitrarily. In particular they may be chosen in such a way that from Eq. (4.30) two sets of equations follow,

$$\begin{aligned}\int_{(V')} \left(\sigma^{ik}_{|i} + f^k\right)\delta u_k dV = 0 \quad \forall V' \subset V \\ \int_{(V')} \left(\mu^{i}_{.k|i} + e_{ilk}\sigma^{il}\right)\delta\psi^k dV = 0 \quad \forall V' \subset V\end{aligned} \tag{4.31}$$

and

$$\begin{aligned}\int_{(\partial V')} \left(\sigma^{ik}n_i - t^k\right)\delta u_k dS = 0 \quad \forall \partial V' \subset \partial V \\ \int_{(\partial V')} \left(\mu^{i}_{.k}n_i - m_k\right)\delta\psi^k dS = 0 \quad \forall \partial V' \subset \partial V\end{aligned} \tag{4.32}$$

These equations result finally to the following set of local stress equilibrium equations,

$$\sigma^{ik}_{|i} + f^k = 0 \quad \forall P(\Theta^i) \in V \tag{4.33}$$

$$\sigma^{ik}_{n_i} = t^k \quad \forall P(\Theta^i) \in \partial V \tag{4.34}$$

and local moment-stress equilibrium equations,

$$\mu^{i}_{\cdot k|i} + e_{ilk}\sigma^{il} = 0 \quad \forall P(\Theta^i) \in V \tag{4.35}$$

$$\mu^{i}_{\cdot k} n_i = m_k \quad \forall P(\Theta^i) \in \partial V \tag{4.36}$$

4.2.2 Cartesian Coordinates

We apply Eqs. (4.33) to (4.36) for a Cartesian description, thus yielding

$$\sigma_{ik} n_i = t_k \tag{4.37}$$

$$\partial_i \sigma_{ik} + f_k = 0 \tag{4.38}$$

and

$$\mu_{ik} n_i = m_k \tag{4.39}$$

$$\partial_i \mu_{ik} + \varepsilon_{imk}\sigma_{im} = 0 \tag{4.40}$$

We observe that the equilibrium Eqs. (4.37) and (4.38) are identical to the ones holding for the Boltzmann continuum and that the equilibrium Eqs. (4.37) and (4.39) introduce the stress-and couple stress tensors as lineal densities for the internal forces in the sense of Cauchy. Due to the moment equilibrium Eq. (4.40), however, the stress tensor in a Cosserat continuum is in general non-symmetric.

As an example we apply the above equilibrium equations for a 2D setting, thus yielding [5] (Fig. 4.2),

$$\begin{aligned} t_1 &= \sigma_{11} n_1 + \sigma_{21} n_2 \\ t_2 &= \sigma_{12} n_1 + \sigma_{22} n_2 \\ m_3 &= \mu_{13} n_1 + \mu_{23} n_2 \end{aligned} \tag{4.41}$$

and (Fig. 4.3)

$$\begin{aligned} &\frac{\partial \sigma_{11}}{\partial x_1} + \frac{\partial \sigma_{21}}{\partial x_2} + f_1 = 0 \\ &\frac{\partial \sigma_{12}}{\partial x_1} + \frac{\partial \sigma_{22}}{\partial x_2} + f_2 = 0 \\ &\frac{\partial \mu_{13}}{\partial x_1} + \frac{\partial \mu_{23}}{\partial x_2} + \sigma_{12} - \sigma_{21} = 0 \end{aligned} \tag{4.42}$$

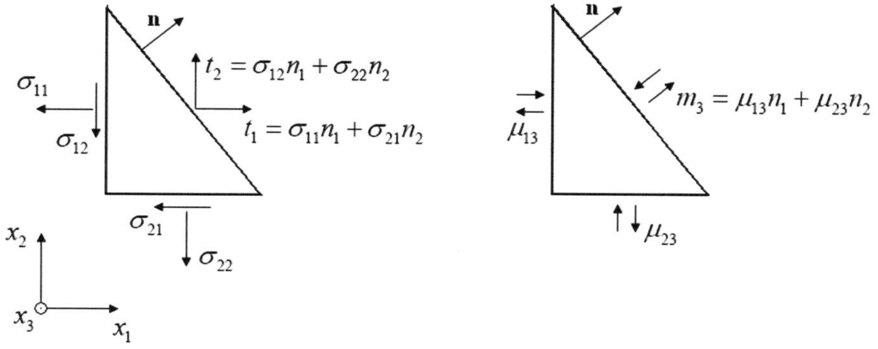

Fig. 4.2 Stress and couple stress in the sense of Cauchy in 2D

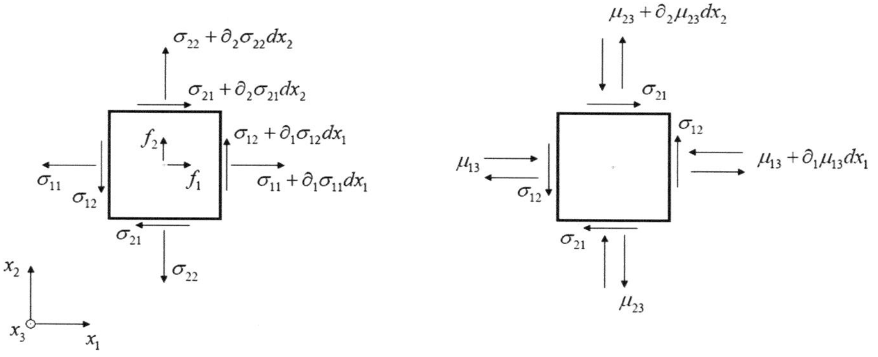

Fig. 4.3 Stress and moment stress equilibrium in 2D

4.2.3 *Exercises: Special Orthogonal Curvilinear Coordinates*

4.2.3.1 Polar Cylindrical Coordinates

Prove that the equilibrium equations for a Cosserat continuum in terms of physical components in polar, cylindrical coordinates are the following:

$$\begin{aligned}
&\frac{\partial\sigma_{rr}}{\partial r} + \frac{1}{r}\frac{\partial\sigma_{\theta r}}{\partial\theta} + \frac{1}{r}(\sigma_{rr} - \sigma_{\theta\theta}) + \frac{\partial\sigma_{zr}}{\partial z} + f_r = 0\\
&\frac{\partial\sigma_{r\theta}}{\partial r} + \frac{1}{r}\frac{\partial\sigma_{\theta\theta}}{\partial\theta} + \frac{1}{r}(\sigma_{r\theta} + \sigma_{\theta r}) + \frac{\partial\sigma_{z\theta}}{\partial z} + f_\theta = 0\\
&\frac{\partial\sigma_{rz}}{\partial r} + \frac{1}{r}\frac{\partial\sigma_{\theta z}}{\partial\theta} + \frac{1}{r}\sigma_{rz} + \frac{\partial\sigma_{zz}}{\partial z} + f_z = 0
\end{aligned} \tag{4.43}$$

and

$$
\begin{aligned}
&\frac{\partial \mu_{rr}}{\partial r}+\frac{1}{r}\frac{\partial \mu_{\theta r}}{\partial \theta}+\frac{\partial \mu_{zr}}{\partial z}+\frac{1}{r}(\mu_{rr}-\mu_{\theta\theta})+\sigma_{\theta z}-\sigma_{z\theta}+\Phi_r=0\\
&\frac{\partial \mu_{r\theta}}{\partial r}+\frac{1}{r}\frac{\partial \mu_{\theta\theta}}{\partial \theta}+\frac{\partial \mu_{z\theta}}{\partial z}+\frac{1}{r}(\mu_{r\theta}+\mu_{\theta r})+\sigma_{zr}-\sigma_{rz}+\Phi_\theta=0\\
&\frac{\partial \mu_{rz}}{\partial r}+\frac{1}{r}\frac{\partial \mu_{\theta z}}{\partial \theta}+\frac{\partial \mu_{zz}}{\partial z}+\frac{1}{r}\mu_{rz}+\sigma_{r\theta}-\sigma_{\theta r}+\Phi_z=0
\end{aligned}
\tag{4.44}
$$

In the above equations with Φ_r, Φ_θ and Φ_z we denote the components of body couples.

4.2.3.2 Polar Spherical Coordinates

Prove that the equilibrium equations for a Cosserat continuum in terms of physical components in polar, cylindrical coordinates are the following:

$$
\begin{aligned}
&\frac{\partial \sigma_{rr}}{\partial r}+\frac{1}{r}\frac{\partial \sigma_{\theta r}}{\partial \theta}+\frac{1}{r\sin\theta}\frac{\partial \sigma_{\phi r}}{\partial \phi}+\frac{\cot\theta}{r}\sigma_{\theta r}+\frac{1}{r}\left(2\sigma_{rr}-\sigma_{\phi\phi}-\sigma_{\theta\theta}\right)+f_r=0\\
&\frac{\partial \sigma_{r\theta}}{\partial r}+\frac{1}{r}\frac{\partial \sigma_{\theta\theta}}{\partial \theta}+\frac{1}{r\sin\theta}\frac{\partial \sigma_{\phi\theta}}{\partial \phi}+\frac{1}{r}(2\sigma_{r\theta}+\sigma_{\theta r})+\frac{\cot\theta}{r}\left(\sigma_{\theta\theta}-\sigma_{\phi\phi}\right)+f_\theta=0\\
&\frac{\partial \sigma_{r\phi}}{\partial r}+\frac{1}{r}\frac{\partial \sigma_{\theta\phi}}{\partial \theta}+\frac{1}{r\sin\theta}\frac{\partial \sigma_{\phi\phi}}{\partial \phi}+\frac{1}{r}\left(\sigma_{r\phi}+\sigma_{\phi r}\right)+\frac{\cot\theta}{r}\left(\sigma_{\theta\phi}+\sigma_{\phi\theta}\right)+f_\phi=0
\end{aligned}
\tag{4.45}
$$

$$
\begin{aligned}
&\frac{\partial \mu_{rr}}{\partial r}+\frac{1}{r}\frac{\partial \mu_{\theta r}}{\partial \theta}+\frac{1}{r\sin\theta}\frac{\partial \mu_{\phi r}}{\partial \phi}+\frac{\cot\theta}{r}\mu_{\theta r}+\frac{1}{r}\left(2\mu_{rr}-\mu_{\phi\phi}-\mu_{\theta\theta}\right)+\sigma_{\phi\theta}-\sigma_{\theta\phi}+\Phi_r=0\\
&\frac{\partial \mu_{r\theta}}{\partial r}+\frac{1}{r}\frac{\mu\sigma_{\theta\theta}}{\partial \theta}+\frac{1}{r\sin\theta}\frac{\partial \mu_{\phi\theta}}{\partial \phi}+\frac{1}{r}(2\mu_{r\theta}+\mu_{\theta r})+\frac{\cot\theta}{r}\left(\mu_{\theta\theta}-\mu_{\phi\phi}\right)+\sigma_{r\phi}-\sigma_{\phi r}+\Phi_\theta=0\\
&\frac{\partial \mu_{r\phi}}{\partial r}+\frac{1}{r}\frac{\partial \mu_{\theta\phi}}{\partial \theta}+\frac{1}{r\sin\theta}\frac{\partial \mu_{\phi\phi}}{\partial \phi}+\frac{1}{r}\left(2\mu_{r\phi}+\mu_{\phi r}\right)+\frac{\cot\theta}{r}\left(\mu_{\theta\phi}+\mu_{\phi\theta}\right)+\sigma_{\theta r}-\sigma_{r\theta}+\Phi_\phi=0
\end{aligned}
\tag{4.46}
$$

4.3 The Traction Motor

In view of Eqs. (4.34) and (4.36) we consider the traction and the couple that are acting on an infinitesimal oriented surface element $\boldsymbol{n}dS$ located at point $P(\Theta^i)$,

$$
\boldsymbol{t}=t^i\boldsymbol{g}_i=\sigma^{ki}n_k\boldsymbol{g}_i \tag{4.47}
$$

$$\boldsymbol{m} = m_i \boldsymbol{g}^i = \mu^k_{\cdot i} n_k \boldsymbol{g}^i \tag{4.48}$$

For the compact description of the Cosserat continuum statics we introduce the vector compound,

$$\underline{T}(n_j, \Theta^j) = \begin{pmatrix} \boldsymbol{t}(n_j, \Theta^j) \\ \boldsymbol{m}(n_j, \Theta^{ji}) \end{pmatrix} = \begin{pmatrix} t^i(n_j, \Theta^j)\boldsymbol{g}_i \\ m_i(n_j, \Theta^j)\boldsymbol{g}^i \end{pmatrix} = \begin{pmatrix} \sigma^{ki}(\Theta^j) n_k \boldsymbol{g}_i \\ \mu^k_{\cdot i}(\Theta^j) n_k \boldsymbol{g}^i \end{pmatrix} \tag{4.49}$$

This prompts to introduce the stress compound

$$\underline{\underline{\Sigma}} = \begin{pmatrix} \sigma^{ki} \boldsymbol{g}_k \otimes \boldsymbol{g}_i \\ \mu^k_{\cdot i} \boldsymbol{g}_k \otimes \boldsymbol{g}^i \end{pmatrix} \tag{4.50}$$

Such that,

$$\underline{T} = \underline{\underline{\Sigma}}^T \cdot \boldsymbol{n} \tag{4.51}$$

This is true, because we recall that the definition of the inner product of the tensor product with a vector obeys the relations [6],

$$(\boldsymbol{a} \otimes \boldsymbol{b}) \cdot \boldsymbol{c} = \boldsymbol{a}(\boldsymbol{b} \cdot \boldsymbol{c}), \quad \boldsymbol{a} \cdot (\boldsymbol{b} \otimes \boldsymbol{c}) = (\boldsymbol{a} \cdot \boldsymbol{b})c \tag{4.52}$$

Thus indeed,

$$\begin{aligned} \underline{\underline{\Sigma}}^T \cdot \boldsymbol{n} &= \begin{pmatrix} \sigma^{ki} \boldsymbol{g}_i \otimes \boldsymbol{g}_k \\ \mu^k_{\cdot i} \boldsymbol{g}^i \otimes \boldsymbol{g}_k \end{pmatrix} \cdot n_l \boldsymbol{g}^l \\ &= \begin{pmatrix} \sigma^{ki} n_l (\boldsymbol{g}_i \otimes \boldsymbol{g}_k) \cdot \boldsymbol{g}^l \\ \mu^k_{\cdot i} n_l (\boldsymbol{g}^i \otimes \boldsymbol{g}_k) \cdot \boldsymbol{g}^l \end{pmatrix} = \begin{pmatrix} \sigma^{ki} n_l \boldsymbol{g}_i \delta^l_k \\ \mu^k_{\cdot i} n_l \boldsymbol{g}^i \delta^l_k \end{pmatrix} = \begin{pmatrix} \sigma^{ki} n_k \boldsymbol{g}_i \\ \mu^k_{\cdot i} n_k \boldsymbol{g}^i \end{pmatrix} = \underline{T} \end{aligned} \tag{4.53}$$

where δ^l_k is the only Kronecker delta,

$$\boldsymbol{g}_k \cdot \boldsymbol{g}^l = \delta^l_k = \begin{cases} 1 & if : k = l \\ 0 & if : k \neq l \end{cases} \tag{4.54}$$

We consider now the virtual displacement and deformation compounds,

$$\delta \bar{K} = \begin{pmatrix} \delta \psi^i \boldsymbol{g}_i \\ \delta u_i \boldsymbol{g}^i \end{pmatrix} \Rightarrow \delta \bar{\bar{H}} = Grad \delta \bar{\text{K}} := \begin{pmatrix} \psi^i_{\cdot|k} \boldsymbol{g}_i \otimes \boldsymbol{g}^k \\ (u_{i|k} + e_{ikl}\psi^l) \boldsymbol{g}^i \otimes \boldsymbol{g}^k \end{pmatrix} = \begin{pmatrix} \delta \kappa^{\cdot i}_k \boldsymbol{g}^k \otimes \boldsymbol{g}_i \\ \delta \gamma_{ki} \boldsymbol{g}^k \otimes \boldsymbol{g}^i \end{pmatrix} \tag{4.55}$$

We observe that the above defined invariant virtual work of internal forces, Eq. (4.3), can be written as a von Misses scalar product of the related static and kinematic tensor compounds,

$$\begin{aligned}
\delta w^{(\mathrm{int})} &= \underline{\underline{\Sigma}} \circ \delta\overline{\overline{H}} \\
&= \begin{pmatrix} \sigma^{ki}\boldsymbol{g}_k \otimes \boldsymbol{g}_i \\ \mu^{k}_{\cdot i}\boldsymbol{g}_k \otimes \boldsymbol{g}^i \end{pmatrix} \circ \begin{pmatrix} \delta\kappa_l^{\cdot j}\boldsymbol{g}^l \otimes \boldsymbol{g}_j \\ \delta\gamma_{lj}\boldsymbol{g}^l \otimes \boldsymbol{g}^j \end{pmatrix} \\
&:= \sigma^{ki}\delta\gamma_{lj}(\boldsymbol{g}_k \otimes \boldsymbol{g}_i) : (\boldsymbol{g}^l \otimes \boldsymbol{g}^j) + \mu^{k}_{\cdot i}\delta\kappa_l^{\cdot j}(\boldsymbol{g}_k \otimes \boldsymbol{g}^i) : (\boldsymbol{g}^l \otimes \boldsymbol{g}_j)
\end{aligned} \tag{4.56}$$

We recall that for two second order tensors a scalar product may be defined in the following manner [6],

$$(\boldsymbol{a} \otimes \boldsymbol{b}) : (\boldsymbol{c} \otimes \boldsymbol{d}) = (\boldsymbol{a} \cdot \boldsymbol{c})(\boldsymbol{b} \cdot \boldsymbol{d}) \tag{4.57}$$

With that Eq. (4.56) reduces to Eq. (4.3), since

$$\begin{aligned}
(\boldsymbol{g}_k \otimes \boldsymbol{g}_i) : (\boldsymbol{g}^l \otimes \boldsymbol{g}^j) &= (\boldsymbol{g}_k \cdot \boldsymbol{g}^l)(\boldsymbol{g}_i \cdot \boldsymbol{g}^j) = \delta_k^l \delta_i^j \\
(\boldsymbol{g}_k \otimes \boldsymbol{g}^i) : (\boldsymbol{g}^l \otimes \boldsymbol{g}_j) &= (\boldsymbol{g}_k \cdot \boldsymbol{g}^l)(\boldsymbol{g}^i \cdot \boldsymbol{g}_j) = \delta_k^l \delta_j^i
\end{aligned} \tag{4.58}$$

and with that from Eq. (4.56) we get in fact that,

$$\delta w^{(\mathrm{int})} = \sigma^{ki}\delta\gamma_{lj}\delta_k^l\delta_i^j + \mu^{k}_{\cdot i}\delta\kappa_l^{\cdot j}\delta_k^l\delta_j^i = \sigma^{ki}\delta\gamma_{ki} + \mu^{k}_{\cdot i}\delta\kappa_k^{\cdot i} \tag{4.59}$$

Note also that the above re-defined von Mises scalar product is commutative, since

$$\begin{aligned}
\delta w^{(\mathrm{int})} &= \delta\overline{\overline{H}} \circ \underline{\underline{\Sigma}} \\
&= \begin{pmatrix} \delta\kappa_l^{\cdot j}\boldsymbol{g}^l \otimes \boldsymbol{g}_j \\ \delta\gamma_{lj}\boldsymbol{g}^l \otimes \boldsymbol{g}^j \end{pmatrix} \circ \begin{pmatrix} \sigma^{ki}\boldsymbol{g}_k \otimes \boldsymbol{g}_i \\ \mu^{k}_{\cdot i}\boldsymbol{g}_k \otimes \boldsymbol{g}^i \end{pmatrix} \\
&= \delta\kappa_l^{\cdot j}\mu^{k}_{\cdot i}(\boldsymbol{g}^l \otimes \boldsymbol{g}_j) : (\boldsymbol{g}_k \otimes \boldsymbol{g}^i) + \delta\gamma_{lj}\sigma^{ki}(\boldsymbol{g}^l \otimes \boldsymbol{g}^j) : (\boldsymbol{g}_k \otimes \boldsymbol{g}_i) \\
&= \delta\kappa_l^{\cdot j}\mu^{k}_{\cdot i}\delta_k^l\delta_j^i + \delta\gamma_{lj}\sigma^{ki}\delta_k^l\delta_i^j = \delta\kappa_l^{\cdot j}\mu^{l}_{\cdot j} + \delta\gamma_{lj}\sigma^{lj}
\end{aligned} \tag{4.60}$$

i.e.,

$$\delta w^{(\mathrm{int})} = \underline{\underline{\Sigma}} \circ \delta\overline{\overline{H}} = \delta\overline{\overline{H}} \circ \underline{\underline{\Sigma}} \tag{4.61}$$

If the Cosserat continuum in the vicinity of a point $P(\Theta^k)$ and for surface elements in an arbitrary direction $\boldsymbol{n}$ is behaving like a rigid body, then the traction $\boldsymbol{t}(n_i, \Theta^i)$ is a line vector and $\boldsymbol{m}(n_i, \Theta^i)$ is the corresponding couple that satisfy the following "transport" law [7],

$$\begin{aligned}
&\boldsymbol{t}(n_j, \Theta^j + d\Theta^j) = \boldsymbol{t}(n_j, \Theta^j) \Rightarrow d\boldsymbol{t} = 0 \\
&\boldsymbol{m}(n_j, \Theta^j + d\Theta^j) = \boldsymbol{m}(n_j, \Theta^j) + \boldsymbol{t}(n_j, \Theta^j) \times d\Theta^k\boldsymbol{g}_k \Rightarrow d\boldsymbol{m} = \boldsymbol{t}(n_j, \Theta^j) \times d\Theta^k\boldsymbol{g}_k
\end{aligned} \tag{4.62}$$

Thus,

$$\underline{T}' = \underline{T}(n_j, \Theta^j + d\Theta^j) = \begin{pmatrix} \boldsymbol{t}(n_j, \Theta^j) \\ \boldsymbol{m}(n_j, \Theta^j) + \boldsymbol{t}(n_j, \Theta^j) \times d\Theta^k \boldsymbol{g}_k \end{pmatrix}$$
$$= \begin{pmatrix} t^i(n_j, \Theta^j)\boldsymbol{g}_i \\ (m_i(n_j, \Theta^j) + e_{ikl}t^k(n_j, \Theta^j)d\Theta^l)\boldsymbol{g}^i \end{pmatrix} \tag{4.63}$$

With,

$$\underline{T}' = \begin{pmatrix} t'^i \boldsymbol{g}_i \\ m'_i \boldsymbol{g}^i \end{pmatrix} \tag{4.64}$$

We get

$$\begin{aligned} t'^i &= t^i \\ m'_i &= m_i + e_{ikl}t^k d\Theta^l \end{aligned} \tag{4.65}$$

This means that in case of a rigidified continuum domain the above introduced generalized-traction vector compound $\overline{T}$, consisting of the two vectors $\boldsymbol{t}$ and $\boldsymbol{m}$, is a motor in the sense of von Mises. We call $\overline{T}$ the traction motor. We observe that with Eq. (4.65) the two motors $\underline{T}(\Theta^i)$ and $\underline{T}' = \underline{T}(\Theta^i + d\Theta^i)$ are "equal" or "statically equivalent". This fundamental statical property of Cosserat continua enforces further their application to the mechanics of granular media, if again the single (rigid) grain is seen as the smallest material unit.

In general the differential forms,

$$d\boldsymbol{t} = \boldsymbol{t}(\Theta^i + d\Theta^i) - \boldsymbol{t}(\Theta^i) \tag{4.66}$$

$$d\boldsymbol{m} - \boldsymbol{t}(\Theta^i) \times d\Theta^k \boldsymbol{g}_k = \boldsymbol{m}(\Theta^k + d\Theta^k) - (\boldsymbol{m}(\Theta^i) + \boldsymbol{t}(\Theta^i)d\Theta^k \boldsymbol{g}_k) \tag{4.67}$$

will not vanish. In compact form this is described by the non-vanishing absolute differential vector compound,

$$d\underline{T}(n_i, \Theta^i) = \begin{pmatrix} \boldsymbol{t}(\Theta^i + d\Theta^i) - \boldsymbol{t}(\Theta^i) \\ \boldsymbol{m}(\Theta^i + d\Theta^i) - (\boldsymbol{m}(\Theta^i) + \boldsymbol{t}(\Theta^i) \times d\Theta^k \boldsymbol{g}_k) \end{pmatrix}$$
$$= \begin{pmatrix} d\boldsymbol{t} \\ d\boldsymbol{m} + d\Theta^k \boldsymbol{g}_k \times \boldsymbol{t} \end{pmatrix} \tag{4.68}$$

In analogy to kinematics, we introduce the Pfaffian vector forms

$$\boldsymbol{\sigma}_i d\Theta^i = d\boldsymbol{t} \tag{4.69}$$

$$\boldsymbol{\mu}_i d\Theta^i = d\boldsymbol{m} + d\Theta^k \boldsymbol{g}_k \times \boldsymbol{t} \tag{4.70}$$

These forms define in turn the two vectors,

$$\boldsymbol{\sigma}_i = \boldsymbol{t}_{,i} \tag{4.71}$$

and

$$\boldsymbol{\mu}_i = \boldsymbol{m}_{,i} + \boldsymbol{g}_i \times \boldsymbol{t} \tag{4.72}$$

with,

$$\boldsymbol{t}_{,i} = t^k_{\cdot|i} \boldsymbol{g}_k \tag{4.73}$$

and

$$\boldsymbol{m}_{,i} = m_{k|i} \boldsymbol{g}^k \tag{4.74}$$

Thus in a given direction $\boldsymbol{n}$ we get

$$\boldsymbol{\sigma}_i = \sigma^{mk}_{\cdot\cdot|i} n_m \boldsymbol{g}_k \tag{4.75}$$

and

$$\boldsymbol{\mu}_i = \left(\mu^m_{\cdot k|i} + e_{ilk}\sigma^{ml}\right) n_m \boldsymbol{g}^k \tag{4.76}$$

With

$$d\underline{T}(n_i, \Theta^i) = \begin{pmatrix} \boldsymbol{\sigma}_i \\ \boldsymbol{\mu}_i \end{pmatrix} d\Theta^i = \begin{pmatrix} \sigma^{mk}_{\cdot\cdot|i} \boldsymbol{g}_k \\ \left(\mu^m_{\cdot k|i} + e_{ilk}\sigma^{ml}\right) \boldsymbol{g}^k \end{pmatrix} n_m d\Theta^i \tag{4.77}$$

and according to Kessel [8], we define the divergence of the stress compound as,

$$\underline{\underline{\Sigma}} = \begin{pmatrix} \sigma^{ki} \boldsymbol{g}_k \otimes \boldsymbol{g}_i \\ \mu^k_{\cdot i} \boldsymbol{g}_k \otimes \boldsymbol{g}^i \end{pmatrix} \Rightarrow Div\underline{\underline{\Sigma}} := \begin{pmatrix} \sigma^{ki}_{\cdot\cdot|k} \boldsymbol{g}_i \\ \left(\mu^k_{\cdot i|k} + e_{ikl}\sigma^{kl}\right) \boldsymbol{g}^i \end{pmatrix} \tag{4.78}$$

From Eqs. (4.33), (4.35) and the definition Eq. (4.78), equilibrium is expressed in compact form as,

$$Div\underline{\underline{\Sigma}} = \begin{pmatrix} -f^i \boldsymbol{g}_i \\ \boldsymbol{0} \end{pmatrix} \tag{4.79}$$

Note that for non-equilibrium stress- and couple-stress states we have to abandon the realm of statics and re-formulate the governing equations within the frame of Cosserat continuum dynamics.

4.4 Statical Compatibility Conditions for Stationary Discontinuities

Within the frame of a small deformation theory, we neglect geometric correction terms and the continued equilibrium across a material discontinuity surface is expressed in terms of the material time derivatives of the Cauchy-type stress and couple-stress tensors [4],

$$\left[\dot{\sigma}^{ik}\right]n_i = 0 \tag{4.80}$$

and

$$\left[\dot{\mu}^{i}_{\cdot k}\right]n_i = 0 \tag{4.81}$$

Following standard techniques, further compatibility equations can be derived from the above discussed differential equilibrium equations, as explained in standards texts [4, 9].

References

1. Brand, L. (1940). *Vector and tensor analysis.*Wiley.
2. Besdo, D. (1974). Ein Beitrag zur nichtlinearen Theorie des Cosserat-Kontinuums. *Acta Mechanica, 20,* 105–131.
3. Mühlhaus, H.-B., & Vardoulakis, I. (1987). The thickness of shear bands in granular materials. *Géotechnique, 37,* 271–283.
4. Vardoulakis, I. & Sulem, J. (1955). *Bifurcation analysis in geomechanics.* Blackie Academic & Professional.
5. Schaefer, H. (1962). *Versuch einer Elastizitätstheorie des zweidimensionalen ebenen Cosserat-Kontinuums,* in *Miszellaneen der angewandten Mechanik,* pp. 277–292. Akademie-Verlag: Berlin.
6. Itskov, M. (1965). *Tensor algebra and tensor analysis for engineers.* Springer.
7. Schaefer, H. (1968). *The basic affine connection in a Cosserat Continuum,* In E. Kröner, (Ed.), *Mechanics of generalized continua,* pp. 57–62. Springer.
8. Kessel, S. (1967). *Stress functions and loading singularities for the infinitely extended linear elastic-isotropic Cosserat Continuum.* In *Proceedings of the IUTAM-symposium on the generalized Cosserat continuum and the continuum theory of dislocations with applications.* Stuttgart: Springer.
9. Thomas, T. Y. (1961). *Plastic flow and fracture.* Vol. 2. Academic Press.

Chapter 5
Cosserat Continuum Dynamics

Abstract In this chapter the conservation (balance) laws for mass, linear and angular momentum are presented. It is shown that the Cosserat continuum differs from the Boltzmann continuum only in the angular momentum balance equation. Examples of how momentum balance laws can be applied to cartesian and polar coordinate systems are also presented.

The equations that describe mass balance and balance of linear momentum in a Cosserat continuum are the same as the ones holding for a Boltzmann continuum [1]. The difference between the two types of continua arises while considering the action of the extra dofs of the Cosserat continuum, i.e. in the formulation of the momentum balance- and energy balance equations [2]. For completeness we derive here also the equations that describe balance of mass and balance of linear momentum.

5.1 Balance of Mass

The material particle of the Cosserat continuum is equipped with a linear particle velocity

$$v^i = \frac{Du^i}{Dt} \tag{5.1}$$

where

$$\frac{Du^i}{Dt} = \frac{\partial u^i}{\partial t} + u^i_{\cdot|k} v^k \tag{5.2}$$

We remark that from Eq. (5.2)

$$v^k\left(\delta^i_k - u^i_{\cdot|k}\right) = \frac{\partial u^i}{\partial t} \Rightarrow v^i = \frac{\partial u^i}{\partial t} + O\left(u^i_{\cdot|k}\right) \tag{5.3}$$

I. Vardoulakis, *Cosserat Continuum Mechanics*, Lecture Notes in Applied and Computational Mechanics 87, https://doi.org/10.1007/978-3-319-95156-0_5

This means that within a small deformation theory we have the following approximation,

$$v^i \approx \partial_t u^i; \quad \partial_t \equiv \frac{\partial}{\partial t} \tag{5.4}$$

The mass of the particle is,

$$dm = \rho dV \tag{5.5}$$

where density $\rho(\Theta^i, t)$ is the mass density at the considered point.

The total mass of a body **B** at a given time t is,

$$M(t) = \int_{(V)} \rho \, dV \tag{5.6}$$

Mass balance is expressed by the requirement,

$$\frac{dM}{dt} = 0 \tag{5.7}$$

We recall Reynolds' transport theorem [1],

$$S(t) = \int_{(V)} s(\Theta^k, t) dV \Rightarrow \frac{dS}{dt} = \int_{(V)} \left(\partial_t s + \left(s v^i\right)_{|i}\right) dV \tag{5.8}$$

Thus, from Eqs. (5.6) and (5.7) follows that mass balance is expressed as

$$\int_{(V')} \left(\partial_t \rho + \left(\rho v^i\right)_{|i} = 0\right) dV = 0 \quad \forall V' \subset V \tag{5.9}$$

If we assume that mass balance holds for any subdivision of the considered body, then from Eq. (5.9) we get the following local form for the mass balance equation,

$$\frac{\partial \rho}{\partial t} + \left(\rho v^i\right)_{|i} = 0 \quad \forall P(\Theta^i) \in V \tag{5.10}$$

or,

$$\dot{\rho} = -\rho v^i_{|i} \tag{5.11}$$

where

$$\dot{\rho} \equiv \frac{D\rho}{Dt} = \frac{\partial \rho}{\partial t} + v^k \rho_{|k} \tag{5.12}$$

is the material time derivative of the density and

$$\rho_{|k} = \rho_{,k} \tag{5.13}$$

Remark Note that if Eq. (5.10) holds, then Reynolds' transport theorem, Eq. (5.8), applied for the global quantity

$$S(t) = \int\limits_{(B)} sdm = \int\limits_{(V)} \rho s\, dV \tag{5.14}$$

yields,

$$\frac{dS}{dt} = \int\limits_{(V)} \rho \dot{s} dV \tag{5.15}$$

where $\dot{s}$ is the material time derivative of the specific quantity $s(\Theta^k, t)$.
This is because,

$$\frac{dS}{dt} = \int\limits_{(V)} \left(\partial_t(\rho s) + \left(\rho s v^i\right)_{|i}\right) dV = \int\limits_{(V)} \left(s\left(\partial_t \rho + \left(\rho v^i\right)_{|i}\right) + \rho\left(\partial_t s + v^i s_{|i}\right)\right) dV \tag{5.16}$$

5.2 Balance of Linear Momentum

The total force that is acting on a body **B** at a given time t is,

$$F^i(t) = \int\limits_{(V)} f^i dV + \int\limits_{(\partial V)} t^i dS \tag{5.17}$$

The total linear momentum of the consider body is,

$$\Gamma^i(t) = \int\limits_{(V)} \rho v^i dV \tag{5.18}$$

Balance of linear momentum is expressed as,

$$\frac{d\Gamma^i}{dt} = F^i \tag{5.19}$$

From Reynolds' transport theorem we get that,

$$\frac{d\Gamma^i}{dt} = \frac{d}{dt}\int\limits_{(V)} \rho v^i dV = \int\limits_{(V)} \rho\left(\partial_t v^i + v^k v^i_{\cdot|k}\right) dV \tag{5.20}$$

Thus from Eqs. (5.17) to (5.20) we get,

$$\int\limits_{(V)} \rho\left(\partial_t v^i + v^k v^i_{\cdot|k}\right) dV = \int\limits_{(V)} f^i dV + \int\limits_{(\partial V)} t^i dS \tag{5.21}$$

We assume that the linear momentum balance holds for any subdivision of the considered body. If we apply Eq. (5.21) in particular for the elementary tetrahedron under suitable mathematical restrictions [3] the volume integrals tend to zero and the remaining surface integral yields Cauchy's theorem, Eq. (4.34),

$$\sigma^{ki} n_k = t^i \quad \forall P(\Theta^m) \in \partial V \tag{5.22}$$

From Eqs. (5.21), (5.22) and Gauss' theorem we get

$$\begin{aligned} \int\limits_{(V')} \rho\left(\partial_t v^i + v^i_{\cdot|k} v^k\right) dV &= \int\limits_{(V')} f^i dV + \int\limits_{(\partial V')} \sigma^{ki} n_k dS \\ &= \int\limits_{(V')} f^i dV + \int\limits_{(V')} \sigma^{ki}_{|k} dV \quad \forall V' \subset V \end{aligned} \tag{5.23}$$

We observe that the material time derivative of the velocity coincides with the particle acceleration,

$$a^i = \frac{Dv^i}{Dt} = \partial_t v^i + v^i_{\cdot|k} v^k \tag{5.24}$$

From Eqs. (5.23) and (5.24) we get the dynamic equations,

$$\sigma^{ki}_{|k} + f^i = \rho \frac{Dv^i}{Dt} \tag{5.25}$$

In Cartesian coordinates the above dynamic equations become,

$$\partial_k \sigma_{ki} + f_i = \rho \frac{Dv_i}{Dt} \tag{5.26}$$

We observe that if we assume that the particle acceleration is vanishing, then Eqs. (5.25) and (5.26) reduce to the static equilibrium Eqs. (4.33) and (4.38), respectively.

5.3 Balance of Angular Momentum

The total moment of the forces and couples acting on a body **B** at a given time t is,

$$\boldsymbol{M} = \int\limits_{(\partial V)} \mathbf{R} \times \boldsymbol{t}\, d\boldsymbol{S} + \int\limits_{(V)} \mathbf{R} \times \boldsymbol{f}\, d\boldsymbol{V} + \int\limits_{(\partial V)} \boldsymbol{m}\, dS \tag{5.27}$$

where $\mathbf{R}$ is the position vector.

On the other hand, the total angular momentum is

$$\boldsymbol{L} = \int\limits_{(V)} \mathbf{R} \times (\rho \boldsymbol{v}) d\boldsymbol{V} + \int\limits_{(V)} \rho \boldsymbol{\theta}\, dV \tag{5.28}$$

where $\rho\boldsymbol{\theta}$ is the angular momentum of the spinning polar material point.

Balance of angular momentum is expressed as,

$$\frac{d\boldsymbol{L}}{dt} = \boldsymbol{M} \tag{5.29}$$

Assuming that mass balance is holding, we have that,

$$\frac{d\boldsymbol{L}}{dt} = \frac{d}{dt} \int\limits_{(V)} \rho(\mathbf{R} \times \boldsymbol{v} + \boldsymbol{\theta}) dV = \int\limits_{(V)} \left(\mathbf{R} \times \rho \frac{D\boldsymbol{v}}{D\boldsymbol{t}}\right) dV + \int\limits_{(V)} \rho \frac{D\boldsymbol{\theta}}{Dt} dV \tag{5.30}$$

5.3.1 Cartesian Coordinates

We consider the 1st term on the r.h.s. of Eq. (5.27), and evaluate its components for convenience in a Cartesian description,

$$\begin{aligned} \int\limits_{(\partial V)} (\mathbf{R} \times \boldsymbol{t})_i dS &= \int\limits_{(\partial V)} \varepsilon_{ijk} x_j t_k dS = \int\limits_{(\partial V)} \varepsilon_{ijk} x_j \sigma_{mk} n_m dS \\ &= \int\limits_{(V)} \varepsilon_{ijk} \partial_m \left(x_j \sigma_{mk}\right) dV = \int\limits_{(V)} \varepsilon_{ijk} \left(\delta_{mj} \sigma_{mk} + x_j \partial_m \sigma_{mk}\right) dV \\ &= \int\limits_{(V)} \left(\varepsilon_{ijk} \sigma_{jk} + \varepsilon_{ijk} x_j \partial_m \sigma_{mk}\right) dV \end{aligned} \tag{5.31}$$

Let,

$$\boldsymbol{t}^* = t_i^* \boldsymbol{e}_i,\ t_i^* = \frac{1}{2}\varepsilon_{ijk}\sigma_{jk} \tag{5.32}$$

and with that Eq. (5.27) becomes

$$\begin{aligned} M_i &= \int_{(V)} 2t_i^* dV + \int_{(V)} \varepsilon_{ijk} x_j \partial_m \sigma_{mk} dV + \int_{(V)} \varepsilon_{ijk} x_j f_k dV + \int_{(\partial V)} \mu_{ki} n_k dS \\ &= \int_{(V)} \left(2t_i^* + \partial_k \mu_{ki}\right) dV + \int_{(V)} \left(\varepsilon_{ijk} x_j (\partial_m \sigma_{mk} + f_k)\right) dV \end{aligned} \tag{5.33}$$

By combining Eqs. (5.29), (5.30) and (5.33) we obtain

$$\int_{(V)} \rho \frac{D\theta_i}{Dt} dV = \int_{(V)} \left(2t_i^* + \partial_k \mu_{ki}\right) dV + \int_{(V)} \left(\varepsilon_{ijk} x_j \left(\partial_m \sigma_{mk} + f_k - \rho \frac{Dv_k}{Dt}\right)\right) dV \tag{5.34}$$

If we assume that balance of linear momentum holds, then the last term on the r. h.s. of Eq. (5.34) is vanishing, thus yielding,

$$\int_{(V)} \rho \frac{D\theta_i}{Dt} dV = \int_{(V)} \left(2t_i^* + \partial_k \mu_{ki}\right) dV \tag{5.35}$$

The local form of Eq. (5.35) is

$$\partial_k \mu_{ki} + \varepsilon_{ikl}\sigma_{kl} = \rho \frac{D\theta_i}{Dt} \tag{5.36}$$

cf. Eq. (4.40).

5.3.2 Exercise: General Curvilinear Coordinates

Prove that in curvilinear coordinates Eq. (5.36) becomes,

$$\mu^{i}_{\cdot k|i} + e_{ilk}\sigma^{il} = \rho \frac{D\theta_k}{Dt} \tag{5.37}$$

Proof From

$$M^i = \int_{(\partial V)} e_{ijk} x^j t^k dS + \int_{(V)} e_{ijk} x^j f^k dV + \int_{(\partial V)} m_i dS + \int_{(V)} \Phi_i dV \tag{5.38}$$

where for completeness we considered the action of body couples as well. Thus,

$$\begin{aligned} M^i &= \int_{(\partial V)} e_{ijk} x^j \sigma^{lk} n_l dS + \int_{(V)} e_{ijk} x^j f^k dV + \int_{(\partial V)} \mu^{\cdot j}_{i} n_j dS + \int_{(V)} \Phi_i dV \\ &= \int_{(V)} \left(\mu^{\cdot j}_{\cdot i|j} + e_{\cdot ijk} \sigma^{jk} + \Phi_i + e_{ijk} x^j \left(\sigma^{lk\cdot}_{\cdot\cdot|l} + f^k \right) \right) dV \end{aligned} \tag{5.39}$$

and

$$L^i = \int_{(V)} e_{ijk} x^j \rho v^k dV + \int_{(V)} \rho \theta_i dV \tag{5.40}$$

Then from Eq. (5.29) we get,

$$\int_{(V)} \left(\mu^{\cdot j}_{\cdot i|j} + e_{\cdot ijk} \sigma^{jk} + \Phi_i - \rho \frac{D\theta_i}{Dt} + e_{ijk} x^j \left(\sigma^{lk\cdot}_{\cdot\cdot|l} + f^k - \rho \frac{Dv^k}{Dt} \right) \right) dV = 0 \tag{5.41}$$

In view of Eq. (5.25) we get Eq. (5.37).

q.e.d.

5.4 The Micro-morphic Continuum

For the identification of the angular momentum of the spinning polar material particle we follow a demonstration by Becker and Bürger [2] by resorting to the so-called micro-morphic continuum interpretation. The term "*micro-morphic*" was introduced by Eringen [4]. A volume element of a micro-morphic medium consists of micro-elements which undergo micro-motions and micro-deformations. Note that micro-polar media are a subclass, in which the micro-elements behave like rigid bodies.

In this case we assign to the material polar particle (or macro-particle) of the continuum the average properties of a *Representative Elementary Volume* (REV); that is of an assembly of sub-particles, as shown in Fig. 5.1. The (REV) may consist of N sub-particles (or micro-particles).

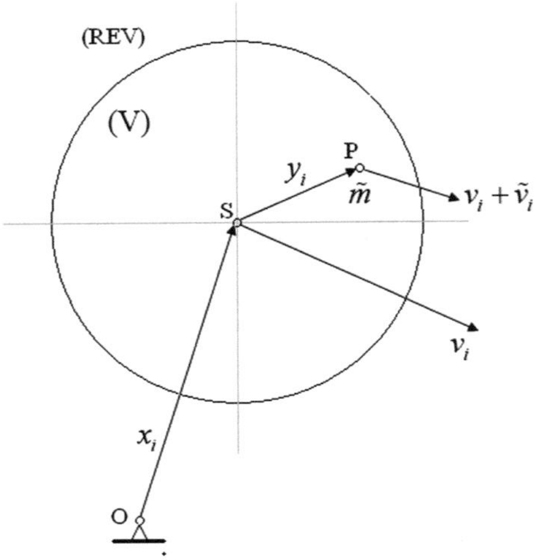

Fig. 5.1 The microstructure of an (REV) with sub-particles sharing a rigid-body motion

We use here Cartesian notation. The spatial position of the polar macro-particle is identified with the position of the center of mass $S(x_i)$ of the sub-particles in the (REV). The velocity v_i of the center of mass $S(x_i)$ is defined as the velocity of the particle itself,

$$v_i = \dot{x}_i \tag{5.42}$$

The sub-particle at position $P(x_i + y_i)$ has the mass $\tilde{m} = \rho_p \widetilde{V}$, where ρ_p is the sub-particle mass density and $\widetilde{V}$ is its volume.

The sub-particle has a velocity that is composed of the velocity of the center of mass and the deviation from that,

$$v_i(P) = v_i(S) + \tilde{v}_i \tag{5.43}$$

The total mass of the macro-particle is the sum of the masses of its constituents,

$$m = \sum_N \tilde{m}_N \tag{5.44}$$

The linear- and angular momentum of the macro-particle are computed as follows,

$$i_i = m v_i \tag{5.45}$$

$$d_k = m \varepsilon_{ijk} x_i v_j + \sum_N \left(\tilde{m} \varepsilon_{ijk} y_i \tilde{v}_j \right)_N \tag{5.46}$$

The volume of the material (REV) is V and the total volume of the sub-particles inside the (REV) is

$$V_p = \sum_N \tilde{V}_N \tag{5.47}$$

The volume fraction

$$\phi = \frac{V - V_p}{V} = 1 - \frac{\sum_N \tilde{V}_N}{V} \tag{5.48}$$

is the porosity of the (REV). The density of the macro-particle is[1]

$$\rho = \frac{\sum_N \tilde{m}_N}{V} = \frac{\sum_N \tilde{m}_N}{\sum_N \tilde{V}_N} \frac{\sum_N \tilde{V}_N}{V} = (1 - \phi)\rho_p \tag{5.49}$$

Similarly we introduce the linear momentum of the macro-particle,

$$s_i = \frac{m v_i}{V} = \rho v_i \tag{5.50}$$

and its angular momentum

$$D_k = \frac{d_k}{V} = \rho \varepsilon_{ijk} x_i v_j + \frac{1}{V} \sum_N \left(\tilde{m} \varepsilon_{ijk} y_i \tilde{v}_j \right)_N \tag{5.51}$$

The relative velocity $\tilde{v}_i$ of the sub-particle at position $P(x_i + y_i)$ with respect to the center of mass $S(x_i)$ is assumed to be a function of its position inside the (REV) and of time. We expand this function in a Taylor series in the vicinity of the center of mass $S(x_i)$ of the REV by setting

$$\tilde{v}_i \approx v_{ij}(t) y_j + v_{ijk}(t) y_j y_k + \cdots \tag{5.52}$$

We can develop a special theory, if we consider only the linear term in the series expansion, Eq. (5.52),

$$\tilde{v}_i \approx v_{ij}(t) y_j \tag{5.53}$$

This assumption is interpreted as a statement for local homogeneity of the micro-deformation; i.e. of the deformation inside the (REV). In this case from Eqs. (5.51) and (5.53) we get,

[1]If the particles consist of different substances then we should replace ρ_p with an average particle density $<\rho_p>$ in Eq. (5.49).

$$D_k = \rho \varepsilon_{ijk} x_i v_j(t) + \varepsilon_{ijk} J_{il} v_{jl}(t) \tag{5.54}$$

where J_{jl} is the inertia tensor of the (REV) with respect to its center of mass,

$$J_{il} = \frac{1}{V} \sum_N (\tilde{m} y_i y_l)_N = \rho_p \frac{1}{V} \sum_N \left(y_i y_l \widetilde{V} \right)_N \tag{5.55}$$

For simplicity we assume that on the (REV) only volume external forces are acting. In this case the moment per unit volume of external forces acting on the (REV) is,

$$\mu_k = \varepsilon_{ijk} x_i f_j + \varepsilon_{ijk} f_{ji} \tag{5.56}$$

where,

$$\begin{aligned} f_i &= \frac{1}{V} \sum_N \left(\tilde{f}_i \widetilde{V} \right)_N \\ f_{ij} &= \frac{1}{V} \sum_N \left(\tilde{f}_i \widetilde{V} y_j \right)_N \end{aligned} \tag{5.57}$$

If we integrate Eq. (5.56) over the volume of the continuum body **B**, we get the expression for the total moment of body forces acting on **B**,

$$M_k^{(b.f.)} = \int_{(V)} \mu_k dV = \int_{(V)} \varepsilon_{ijk} x_i f_j dV + \int_{(V)} \varepsilon_{ijk} f_{ji} dV \tag{5.58}$$

In view of Eq. (5.27) we recognize the 1st term on the r.h.s. of Eq. (5.58), as the moment of body forces. The 2nd term is the contribution of body-couples, that were systematically ignored in the previous derivations, since there was no real motivation to introduce such body-couples until this point in the demonstration. Thus we introduce here body couples,

$$\Phi_k = \varepsilon_{ijk} f_{ji} \tag{5.59}$$

and Eq. (5.58) becomes,

$$M_k^{(b.f.)} = \int_{(V)} \varepsilon_{ijk} x_i f_j dV + \int_{(V)} \Phi_k dV \tag{5.60}$$

With this background we may re-write the linear- and angular momentum equations for the considered special micro-morphic continuum; these are,

$$\frac{d}{dt}\int_{(V)} \rho v_i dV = \int_{(V)} f_i dV + \cdots \tag{5.61}$$

$$\frac{d}{dt}\int_{(V)} D_k dV = \int_{(V)} \varepsilon_{ijk} x_i f_j dV + \int_{(V)} \Phi_k dV + \cdots \tag{5.62}$$

where the dots stand for the actions of surface tractions and surface couples.

Equation (5.62) with (5.54) becomes,

$$\frac{d}{dt}\int_{(V)} \left(\rho \varepsilon_{ijk} x_i v_j + \varepsilon_{ijk} J_{il} v_{jl}\right) dV = \int_{(V)} \varepsilon_{ijk} x_i f_j dV + \int_{(V)} \Phi_k dV + \cdots \tag{5.63}$$

The balance equations for the Cosserat continuum can be derived from Eq. (5.63) if we set,

$$v_{ij} = \dot{\psi}_{ij} = -\varepsilon_{ijk} w_k \tag{5.64}$$

where

$$w_k = \frac{D\psi_k}{Dt} \approx \partial_t \psi_k \tag{5.65}$$

This assumption means that the considered macro-element is a swarm of v sub-particles that they all share a rigid body motion: The center of mass of these sub-particles is translated by the velocity v_i and at the same time all the sub-particles rotate around an instantaneous axis with director n_k and have an angular velocity w, such that,

$$w_k = n_k w \tag{5.66}$$

Thus, the spin and the velocity of a sub-particle inside the (REV) is given by the von Mises motor that characterizes the rigid-body motion of the sub-particles,

$$\begin{pmatrix} \boldsymbol{w}(P) = \boldsymbol{w}(S) \\ \boldsymbol{v}(P) = \boldsymbol{v}(S) + \boldsymbol{w} \times (\mathbf{R}_P - \mathbf{R}_s) \end{pmatrix} = \begin{pmatrix} w_i \\ v_i + \varepsilon_{ikj} w_k y_j \end{pmatrix} \tag{5.67}$$

In this case we have that

$$\varepsilon_{ijk} J_{il} v_{jl} = \varepsilon_{ijk} J_{il}\left(-\varepsilon_{jlm} w_m\right) = J_{km} w_m = \rho \theta_k \tag{5.68}$$

With these results we return to the momentum balance Eq. (5.35), that is written now as follows,

$$\int_{(V)} \rho J_{im} \frac{Dw_m}{Dt} dV = \int_{(V)} \left(t_i^* + \partial_k \mu_{ki} + \Phi_i\right) dV \tag{5.69}$$

Its local form reads,

$$\partial_p \mu_{pk} + \varepsilon_{kpq} \sigma_{pq} + \Phi_k = \rho J_{km} \frac{Dw_m}{Dt} \tag{5.70}$$

In general curvilinear coordinates the dynamic Eq. (5.70) takes the following form,

$$\mu^{p}_{\cdot k|p} + e_{kpq} \sigma^{pq} + \Phi_k = \rho J_{km} \frac{Dw^m}{Dt} \tag{5.71}$$

5.5 Exercises: Dynamic Equations in Plane Problems

5.5.1 Cartesian Coordinates

Prove that the dynamic equations in-plane Cartesian coordinates are (Fig. 5.2):

$$\begin{aligned} \frac{\partial \sigma_{11}}{\partial x_1} + \frac{\partial \sigma_{21}}{\partial x_2} &= \rho \frac{Dv_1}{Dt} \\ \frac{\partial \sigma_{12}}{\partial x_1} + \frac{\partial \sigma_{22}}{\partial x_2} &= \rho \frac{Dv_2}{Dt} \end{aligned} \tag{5.72}$$

$$\frac{\partial \mu_{13}}{\partial x_1} + \frac{\partial \mu_{23}}{\partial x_2} + \sigma_{12} - \sigma_{21} = \rho J \frac{Dw_3}{Dt} \tag{5.73}$$

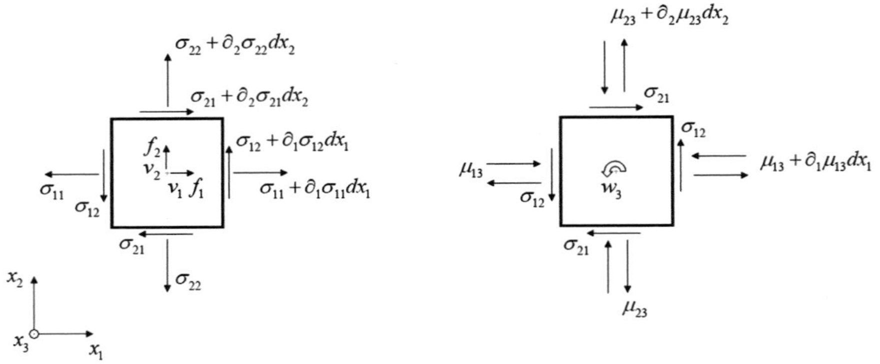

Fig. 5.2 Dynamic equilibrium in a Cosserat medium in Cartesian coordinates

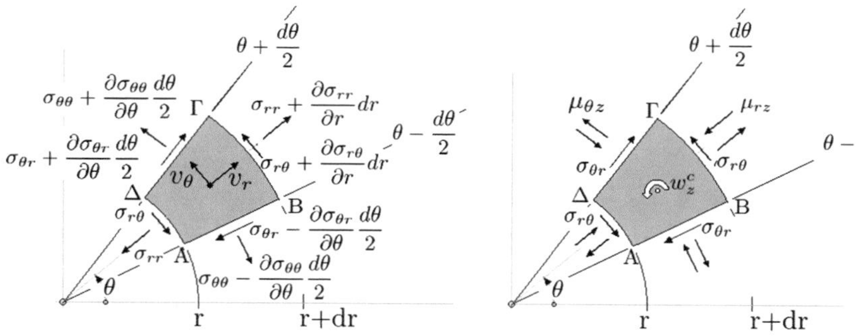

Fig. 5.3 Dynamic equilibrium in a Cosserat medium

where the assumption is made that the micro-inertia tensor is isotropic,

$$J_{ij} = J\delta_{ij} \tag{5.74}$$

5.5.2 Plane, Polar Coordinates

Prove that in the absence of body forces and body couples the dynamic equations in physical components and in plane polar coordinates are (Fig. 5.3),

$$\begin{aligned} \frac{\partial \sigma_{rr}}{\partial r} + \frac{1}{r}\frac{\partial \sigma_{\theta r}}{\partial \theta} + \frac{1}{r}(\sigma_{rr} - \sigma_{\theta\theta}) &= \rho \frac{Dv_r}{Dt} \\ \frac{\partial \sigma_{r\theta}}{\partial r} + \frac{1}{r}\frac{\partial \sigma_{\theta\theta}}{\partial \theta} + \frac{1}{r}(\sigma_{r\theta} + \sigma_{\theta r}) &= \rho \frac{Dv_\theta}{Dt} \end{aligned} \tag{5.75}$$

$$\frac{\partial \mu_{rz}}{\partial r} + \frac{1}{r}\frac{\partial \mu_{\theta z}}{\partial \theta} + \frac{1}{r}\mu_{rz} + \sigma_{r\theta} - \sigma_{\theta r} = \rho J \frac{Dw_z}{Dt} \tag{5.76}$$

References

1. Vardoulakis, I., & Sulem, J. (1995). *Bifurcation analysis in geomechanics*. Blackie Academic & Professional.
2. Becker, E., & Bürger, W. (1975). Kontinuumsmechanik. In H. Görtler (Ed.), *Leitfäden der angewandten Mathematik und Mechanik* (p. 229). Springer.
3. Gurtin, M. E., & Martins, L. C. (1976). Cauchy's theorem in classical physics. *Archive for Rational Mechanics and Analysis, 60,* 305–324.
4. Eringen, A. C. (1965). Theory of micropolar continua. In *Proceedings of the 9th Midwestern mechanics conference*. Madison: Wiley.

Chapter 6
Cosserat Continuum Energetics

Abstract In this chapter the energy and entropy balance laws for a Cosserat continuum are presented. It is shown that the higher-order continuum introduces additional terms in the local dissipation of a system.

6.1 Energy Balance Equation

The total energy $E(t)$ of a continuum body **B** is split into two parts: one part that depends on the position of the observer, that is called the kinetic energy $K(t)$ of the considered body, and another part that does not depend on the observer, called the internal energy $U(t)$,

$$E(t) = K(t) + U(t) \tag{6.1}$$

The kinetic energy of a Cosserat continuum consists of the contribution that is due to the translationary motion of the particles,

$$\frac{1}{2} v^k v_k dm \tag{6.2}$$

and of the contribution that is due to their spin,

$$\frac{1}{2} w^k \theta_k dm \tag{6.3}$$

Thus the total kinetic energy is computed as,

$$K(t) = \int_{(V)} \rho \left(\frac{1}{2} v^k v_k + \frac{1}{2} w^k \theta_k \right) dV \tag{6.4}$$

I. Vardoulakis, *Cosserat Continuum Mechanics*, Lecture Notes in Applied and Computational Mechanics 87, https://doi.org/10.1007/978-3-319-95156-0_6

The internal energy is given by means of a specific internal energy density function, $e(\Theta^i, t)$,

$$U(t) = \int_{(B)} edm = \int_{(V)} \rho edV \tag{6.5}$$

The 1st Law of Thermodynamics requires that the change of the total energy of a material body **B** is due to two factors: (a) the power $W^{(ext)}$ of all external forces acting on **B** in the current configuration, and (b) the non-mechanical energy Q, which is supplied per unit time to **B** from the exterior domain; i.e.:

$$\frac{dE}{dt} = W^{(ext)} + Q \tag{6.6}$$

By eliminating dE/dt from Eqs. (6.1) and (6.6) we arrive to the *fundamental energy balance equation*

$$\frac{dU}{dt} + \frac{dK}{dt} = W^{(ext)} + Q \tag{6.7}$$

According to Truesdell and Toupin [1] the first formulation of the Energy Balance Law, Eq. (6.7), is due to Duhem [2].

The work of external forces is computed as follows,

$$W^{(ext)} = \int_{(\partial V)} \left(t^k v_k + m_k w^k\right) dS + \int_{(V)} \left(f^k v_k + \Phi_k w^k\right) dV \tag{6.8}$$

The influx of energy in the form of heat flow is defined through a heat-flow vector $q^i(\Theta^k, t)$, that is taken positive if heat flows into the considered body,

$$Q = -\int_{(\partial V)} q^k n_k dS \tag{6.9}$$

We introduce into Eq. (6.8) the stress- and couple-stress tensors, according to Eqs. (4.34) and (4.36),

$$W^{(ext)} = \int_{(\partial V)} \left(\sigma^{ik} v_k + \mu^i_{.k} w^k\right) n_i dS + \int_{(V)} \left(f^k v_k + \Phi_k w^k\right) dV \tag{6.10}$$

and we apply Gauss' theorem,

$$W^{(ext)} = \int\limits_{(V)} \left(\sigma^{ik}_{|i} + f^k\right) v_k dV + \int\limits_{(V)} \left(\mu^i_{\cdot k|i} + \Phi_k\right) w^k dV + \int\limits_{(V)} \left(\sigma^{ik} v_{k|i} + \mu^i_{\cdot k} w^k_{\cdot|i}\right) dV \tag{6.11}$$

Similarly from Eq. (6.9) we get

$$Q = -\int\limits_{(V)} q^k_{\cdot|k} dV \tag{6.12}$$

The l.h.s. of Eq. (6.6) is computed by means of Eqs. (6.1) to (6.5) and Reynolds' transport theorem

$$\frac{dE}{dt} = \int\limits_{(V)} \rho \frac{1}{2}\frac{D}{Dt}\left(v^k v_k\right) dV + \int\limits_{(V)} \rho \frac{1}{2}\frac{D}{Dt}\left(w^k \theta_k\right) dV + \int\limits_{(V)} \rho \frac{de}{dt} dV \tag{6.13}$$

We remark that the following relations hold,

$$\frac{1}{2}\frac{D}{Dt}\left(v^k v_k\right) = \frac{1}{2}\frac{D}{Dt}\left(v^k g_{kl} v^l\right) = g_{kl} v^l \frac{Dv^k}{Dt} = v_k \frac{Dv^k}{Dt} \tag{6.14}$$

$$\frac{D\theta_k}{Dt} w^k = \theta_k \frac{Dw^k}{Dt} \tag{6.15}$$

the latter will be demonstrated below. Thus,

$$\frac{1}{2}\frac{D}{Dt}\left(\theta_k w^k\right) = \frac{1}{2}\left(\frac{D\theta_k}{Dt} w^k + \theta_k \frac{Dw^k}{Dt}\right) = w^k \frac{D\theta_k}{Dt} \tag{6.16}$$

By combining Eqs. (6.6), (6.11) with (6.16) we get

$$\begin{aligned} &\int\limits_{(V)} \rho \frac{Dv^k}{Dt} v_k dV + \int\limits_{(V)} \rho \frac{D\theta_k}{Dt} w^k dV + \int\limits_{(V)} \rho \frac{De}{Dt} dV \\ &= \int\limits_{(V)} \left(\sigma^{ik}_{|i} + f^k\right) v_k dV + \int\limits_{(V)} \left(\mu^i_{\cdot k|i} + \Phi_k\right) w^k dV + \int\limits_{(V)} \left(\sigma^{ik} v_{k|i} + \mu^i_{\cdot k} w^k_{\cdot|i}\right) dV - \int\limits_{(V)} q^k_{\cdot|k} dV \end{aligned} \tag{6.17}$$

or

$$\int\limits_{(V)} \rho \frac{De}{Dt} dV = \int\limits_{(V)} \left(\sigma^{ik}_{|i} + f^k - \rho \frac{Dv^k}{Dt} \right) v_k dV + \int\limits_{(V)} \left(\mu^{i}_{\cdot k|i} + \Phi_k - \rho \frac{D\theta_k}{Dt} \right) w^k dV$$
$$+ \int\limits_{(V)} \left(\sigma^{ik} v_{k|i} + \mu^{i}_{\cdot k} w^{k}_{\cdot|i} \right) dV - \int\limits_{(V)} q^{k}_{\cdot|k} dV \tag{6.18}$$

We assume that linear- and angular momentum balance equations, Eqs. (5.25) and (5.37), are holding. In this case we get from Eq. (6.18),

$$\int\limits_{(V)} \rho \frac{De}{Dt} dV = \int\limits_{(V)} \left(\sigma^{ik} \left(v_{k|i} - e_{ikm} w^m \right) + \mu^{i}_{\cdot k} w^{k}_{\cdot|i} - q^{k}_{\cdot|k} \right) dV \tag{6.19}$$

In accordance to Eqs. (3.138) and (3.139) we define the *rate of distortion tensor*

$$K_i^{\cdot k} = w^{k}_{\cdot\ |i} \tag{6.20}$$

and the *relative rate of deformation tensor*,

$$\Gamma_{ik} = v_{k|i} + e_{kil} w^l \tag{6.21}$$

With this notation and in accordance to Eq. (4.3) we define the *stress power* in a Cosserat continuum as,

$$P = \sigma^{ik} \Gamma_{ik} + \mu^{i}_{\cdot k} K_i^{\cdot k} \tag{6.22}$$

Notice that in accordance to Eq. (4.14) the stress power can be split as follows,

$$P = \sigma^{(ij)} \Gamma_{(ij)} + 2t_i^* \left(\dot{\omega}^i - w^i \right) + \mu^{i}_{\cdot k} K_i^{\cdot k} \tag{6.23}$$

With this definition, from Eq. (6.19) we obtain the following local form of the energy balance equation,

$$\rho \frac{De}{Dt} = P - q^{k}_{\cdot|k} \tag{6.24}$$

Remark In order to have the above derivation complete we must prove the validity of Eq. (6.15); cf. Becker and Bürger [3]. We use Cartesian coordinates and Eqs. (5.55) and (5.68) as starting points,

$$\theta_k = J^*_{km} w_m \tag{6.25}$$

where J^*_{ij} is the (symmetric) inertia tensor referred the unit of mass

$$J^*_{il} = \frac{1}{\rho} J_{il} = \frac{1}{m} \sum_N (\tilde{m} y_i y_l)_N; \quad J_{ij} = \rho J^*_{il} \tag{6.26}$$

In view of the l.h.s. of Eqs. (6.15), (6.25) and the symmetry of the moment of inertia tensor we get,

$$\begin{aligned} \frac{D\theta_k}{Dt} w_k &= \left(\frac{DJ^*_{km}}{Dt} w_m + J^*_{km} \frac{Dw_m}{Dt} \right) w_k = w_k \frac{DJ^*_{km}}{Dt} w_m + \frac{Dw_m}{Dt} J^*_{km} w_k \\ &= w_k \frac{DJ^*_{km}}{Dt} w_m + \frac{Dw_m}{Dt} J^*_{mk} w_k \end{aligned} \tag{6.27}$$

In a Cosserat continuum the particles move like rigid bodies. This means that for a fixed coordinate system, the inertia tensor is,

$$J^*_{ij}(t) = Q_{in}(t) Q_{jm}(t) J^*_{nm}(0) \tag{6.28}$$

The proper orthogonal tensor $Q_{ij}(t)$ describes the (finite) rotation of the particle between its configuration at time $t = 0$ and time $t > 0$.

We recall that $Q_{ij}(t)$ fulfills the orthogonality conditions,

$$Q_{ik} Q_{il} = \delta_{kl}, \quad Q_{ki} Q_{li} = \delta_{kl} \tag{6.29}$$

Thus,

$$\frac{D}{Dt} J^*_{ij} = \dot{Q}_{in} Q_{jm} J^*_{nm}(0) + Q_{in}(t) \dot{Q}_{jm}(t) J^*_{nm}(0) \tag{6.30}$$

If we take the current configuration as reference configuration, then Eq. (6.30) yields

$$\frac{D}{Dt} J^*_{ij} = \dot{Q}_{in} \delta_{jm} J^*_{nm} + \delta_{in} \dot{Q}_{jm} J^*_{nm} = \dot{Q}_{in} J^*_{nj} + \dot{Q}_{jm} J^*_{im} \tag{6.31}$$

or

$$\frac{D}{Dt} J^*_{ij} = \Omega_{in} J^*_{nj} - J^*_{im} \Omega_{mj} \tag{6.32}$$

where the tensor

$$\Omega_{kl} = \dot{Q}_{kl} \tag{6.33}$$

is antisymmetric and has the angular velocity vector w_k as an axial vector,

$$[\Omega] = \begin{bmatrix} 0 & -w_3 & w_2 \\ w_3 & 0 & -w_1 \\ -w_2 & w_1 & 0 \end{bmatrix} \tag{6.34}$$

or

$$\Omega_{ij} = -\varepsilon_{ijk} w_k \tag{6.35}$$

Thus,

$$\frac{D}{Dt} J^*_{ij} = -\varepsilon_{ink} w_k J^*_{nj} + \varepsilon_{mjk} w_k J^*_{im} \tag{6.36}$$

and

$$\begin{aligned} w_i \frac{DJ^*_{ij}}{Dt} w_j &= w_i \left(-\varepsilon_{imk} w_k J^*_{mj} + \varepsilon_{mjk} w_k J^*_{im} \right) w_j \\ &= \varepsilon_{mik} w_k w_i J^*_{mj} w_j + \varepsilon_{mjk} w_k w_j J^*_{mi} w_i = 2 r_m (\varepsilon_{mik} w_i w_k) = 0 \end{aligned} \tag{6.37}$$

q.e.d.

6.2 Stress Power in Micro-morphic Media

At this point we would like to make a reference to the more general formulation of the stress power that applies for a micro-morphic medium [4]; cf. Sect. 5.4. In this context we define the following kinematic variables:

The rate of *macro-deformation,*

$$D_{ij} = \frac{1}{2} \left(\partial_i v_j + \partial_j v_i \right) \tag{6.38}$$

the *micro-deformation* v_{ij}, the rate of *relative deformation,*

$$\Gamma_{ij} = \partial_i v_k - v_{ij} \tag{6.39}$$

and the rate of *micro-deformation gradient,*

$$\mathrm{K}_{ijk} = \partial_i v_{jk} \tag{6.40}$$

Based on these definitions we define the stress power as follows,

$$P = \tau_{ij} D_{ij} + \alpha_{ij} \Gamma_{ij} + \mu_{ijk} \mathrm{K}_{ijk} \tag{6.41}$$

The tensor τ_{ij} is called the *macro-stress*, the tensor α_{ij} is the *relative stress* and the tensor μ_{ijk} is the *double stress* [5].

We specialize now the micro-deformation so as to correspond to a rigid-body rotation [cf. Eq. (5.64)],

$$v_{ij} = \dot{\psi}_{ij} = -\varepsilon_{ijk} w_k \tag{6.42}$$

and with that,

$$\mathrm{K}_{ijk} = \partial_i v_{jk} = -\varepsilon_{jkl} \partial_i w_l = -\varepsilon_{jkl} \mathrm{K}_{il} \tag{6.43}$$

In that case from Eq. (6.41) we get,

$$\begin{aligned} P &= \tau_{ij} D_{ij} + \alpha_{ij} \Gamma_{ij} + \mu_{ijk} \mathrm{K}_{ijk} \\ &= \tau_{ij} D_{ij} + \alpha_{ij} \left(\partial_i v_j - \dot{\psi}_{ij} \right) - \varepsilon_{jkl} \mu_{ijk} \mathrm{K}_{il} \\ &= \tau_{ij} D_{ij} + \alpha_{ij} \left(D_{ij} + W_{ij} + \varepsilon_{ijk} w_k \right) - \varepsilon_{jkl} \mu_{ijk} \mathrm{K}_{il} \\ &= \left(\tau_{ij} + \alpha_{(ij)} \right) D_{ij} + \alpha_{[ij]} \left(W_{ij} + \varepsilon_{ijk} w_k \right) - \varepsilon_{jkl} \mu_{ijk} \mathrm{K}_{il} \end{aligned} \tag{6.44}$$

where W_{ij} is called the *macro-spin*

$$W_{ij} = \frac{1}{2} \left(\partial_j v_i - \partial_i v_j \right) \tag{6.45}$$

If we compare Eqs. (4.14) and (6.44), we obtain the following identification among the stress fields defined in the micro-morphic and the Cosserat continuum, respectively,

$$\begin{aligned} \sigma_{(ij)} &= \tau_{ij} + \alpha_{(ij)} \\ \sigma_{[ij]} &= \alpha_{[ij]} \\ \mu_{il} &= -\varepsilon_{jkl} \mu_{ijk} \end{aligned} \tag{6.46}$$

If we assume also that,

$$\alpha_{(ij)} = 0 \tag{6.47}$$

this identification allows us to write the stress power for a Cosserat continuum as follows,

$$P = \tau_{ij} D_{ij} + \alpha_{ij} R_{ij} + \mu_{ij} \mathrm{K}_{ij} \tag{6.48}$$

where R_{ij} is the relative spin

$$R_{ij} = W_{ij} + \varepsilon_{ijk} w_k \tag{6.49}$$

This means that in a Cosserat continuum, the macro-stress coincides with the symmetric part of the true (equilibrium) stress and works on the rate of deformation. The relative stress coincides with the antisymmetric part of the true stress and works on the relative spin. Finally, the couple stresses of the micromorphic medium collapse to the true (equilibrium) couple stresses, that work in turn on the rate of distortion.

6.3 Entropy Balance

Let H be the total entropy of the considered material body **B** in the current configuration

$$\mathrm{H}(t) = \int_V \rho s dV \tag{6.50}$$

In Eq. (6.50) $s = s(x_k, t)$ is the specific entropy. Let also $T = T(x_k, t) > 0$ be the absolute temperature. We define further the following quantities: (a) The Helmholtz free energy as that portion of the internal energy, which is available for doing mechanical work at constant temperature

$$f = e - sT \tag{6.51}$$

and (b) the local dissipation function,

$$D_{loc} = P - \rho\left(\frac{Df}{Dt} + s\frac{DT}{Dt}\right) \tag{6.52}$$

With the above definitions the energy balance Eq. (6.24) becomes,

$$\rho\left(T\frac{Ds}{Dt}\right) = -q^{k}_{\cdot|k} + D_{loc} \tag{6.53}$$

This equation is the *balance law for local entropy production*. The entropy balance law, Eq. (6.53), is further worked out by introducing appropriate constitutive assumptions concerning the Helmholtz free energy function, the local dissipation function and the law of heat conduction that are compatible with the underlying Cosserat structure of the medium.

References

1. Truesdell, C., & Toupin, R. A. (1960). The classical field theories. In S. Flügge (Ed.), *Principles of thermodynamics and statistics*. Springer.
2. Duhem, P. Commentaire aux principes de Ia thermodynamique, Premiere partie. *Journal de Mathématiques Pures et Appliquées*, *8* (4) , 269–330.
3. Becker, E., & Bürger, W. (1975). Kontinuumsmechanik. In H. Görtler (Ed.), *Leitfäden der angewandten Mathematik und Mechanik* (vol. 229). Springer.
4. Mindlin, R. D. (1964). Microstructure in linear elasticity. *Archive for Rational Mechanics and Analysis, 16,* 51–78.
5. Vardoulakis, I., & Giannakopoulos, A. E. (2006). An example of double forces taken from structural analysis. *International Journal of Solids and Structures, 43,* 4047–4062.

Chapter 7
Cosserat-Elastic Bodies

Abstract This chapter explores certain classes of materials. It begins with linear elastic bodies, laying down the basic concepts for linear isotropic elasticity. Examples are following, including bending of a beam, annular shear and torsion of a sphere of Cosserat elastic material.

7.1 Linear, Isotropic Cosserat Elasticity

For an elastic Cosserat material that is stressed under isothermal conditions, the energy balance Eq. (6.24) provides the means to compute the rate of the internal energy density function.

In Cartesian coordinates we have,

$$\rho \frac{De}{Dt} = \sigma_{ij}\Gamma_{ij} + \mu_{ij}K_{ij} \tag{7.1}$$

Within the frame of a small-deformation theory we assume that the density remains practically constant,

$$\frac{\rho}{\rho^{(0)}} = \frac{dV}{dV^{(0)}} \Rightarrow \rho = \frac{\rho^{(0)}}{1+\varepsilon_{kk}} \approx \rho^{(0)}(1-\varepsilon_{kk}) \approx \rho^{(0)} \tag{7.2}$$

where $\rho^{(0)}$ is the density of the material in the initial, unstrained configuration.

Thus for small deformations we get,

$$\begin{aligned} \Gamma_{ij} &= \frac{D\gamma_{ij}}{Dt} \approx \partial_t \gamma_{ij} \\ K_{ij} &= \frac{D\kappa_{ij}}{Dt} \approx \partial_t \kappa_{ij} \end{aligned} \tag{7.3}$$

I. Vardoulakis, *Cosserat Continuum Mechanics*, Lecture Notes in Applied and Computational Mechanics 87, https://doi.org/10.1007/978-3-319-95156-0_7

and

$$\rho \frac{De}{Dt} \approx \rho^{(0)} \frac{De}{Dt} \approx \sigma_{ij}\Gamma_{ij} + \mu_{ij}\mathrm{K}_{ij} \tag{7.4}$$

We assume that the elastic energy density,

$$w^{(el)} = \rho^{(0)} e \tag{7.5}$$

is a function of the 18 kinematic variables, γ_{ij} and κ_{ij},

$$w^{(el)} = F\left(\gamma_{ij}, \kappa_{ij}\right) \tag{7.6}$$

Then from Eqs. (7.6) and (7.4) we get that

$$\sigma_{ij} = \frac{\partial w^{(el)}}{\partial \gamma_{ij}}; \quad \mu_{ij} = \frac{\partial w^{(el)}}{\partial \kappa_{ij}} \tag{7.7}$$

Within the frame of linear elasticity, we assume that $w^{(el)} = F\left(\gamma_{ij}, \kappa_{ij}\right)$ is homogeneous of degree 2 with respect to its arguments.

From

$$\sigma_{ij}\gamma_{ij} = \sigma_{(ij)}\gamma_{(ij)} + \sigma_{[ij]}\gamma_{[ij]} \tag{7.8}$$

we get that the elastic strain energy is split into three terms, as

$$w^{(el)} = w_1^{(el)}\left(\gamma_{(ij)}\right) + w_2^{(el)}\left(\gamma_{[ij]}\right) + w_3^{(el)}\left(\kappa_{ij}\right) \tag{7.9}$$

A simple elasticity model arises if we assume that the 1st term on the r.h.s. of Eq. (7.9) reflects Hooke's law for linear, isotropic, elastic materials [1],

$$w_1^{(el)} = \frac{1}{2}\sigma_{(ij)}\varepsilon_{ij} \Rightarrow \tag{7.10}$$

$$w_1^{(el)} = \frac{1}{2}\lambda\varepsilon_{mm}\varepsilon_{nn} + G\varepsilon_{mn}\varepsilon_{mn}; \quad \lambda = \frac{2v}{1-2v}G \tag{7.11}$$

where G and v are the elastic shear modulus and Poisson's ratio, respectively.

Equations (7.10) and (7.11) are yielding,

$$\sigma_{(ij)} = \frac{\partial w^{(el)}}{\partial \varepsilon_{ij}} = \frac{\partial w_1^{(el)}}{\partial \varepsilon_{ij}} = 2G\left(\varepsilon_{ij} + \frac{v}{1-2v}\varepsilon_{kk}\delta_{ij}\right) \tag{7.12}$$

We notice also that both, $\sigma_{[ij]}$ and $\gamma_{[ij]}$, are antisymmetric tensors of 2nd order. Both are possessing axial vectors, say

$$\sigma_{[ij]} = \varepsilon_{ijk} t_k^*; \quad \gamma_{[ij]} = \varepsilon_{ijk} \gamma_k^* \tag{7.13}$$

where

$$\gamma_k^* = \omega_k - \psi_k \tag{7.14}$$

Isotropy calls for proportionality between the axial vector of the antisymmetric stress and the antisymmetric part of the deformation, thus yielding

$$t_i^* = 2\eta_1 G \gamma_i^* \quad (\eta_1 > 0) \Rightarrow \sigma_{[ij]} = 2\eta_1 G \gamma_{[ij]} \tag{7.15}$$

and with that

$$w_2^{(el)} = \frac{1}{2} \sigma_{[ij]} \gamma_{[ij]} = \eta_1 G \gamma_{[ij]} \gamma_{[ij]} = 6\eta_1 G \gamma_k^* \gamma_k^* \tag{7.16}$$

The couple stress tensor and the gradient of the Cosserat rotation ψ_k are also decomposed additively into symmetric and antisymmetric parts,

$$\mu_{ij} = \mu_{(ij)} + \mu_{[ij]}; \quad \kappa_{ij} = \kappa_{(ij)} + \kappa_{[ij]} \tag{7.17}$$

Then the isotropic linear-elastic law for the couple stress reads,

$$\begin{aligned} \mu_{(ij)} &= \frac{\partial w_3^{(el)}}{\partial \kappa_{(ij)}} = G\ell^2 \left(\kappa_{(ij)} + \eta_2 \delta_{ij} \kappa_{kk} \right), \quad \eta_2 > 0 \\ \mu_{[ij]} &= \frac{\partial w_3^{(el)}}{\partial \kappa_{[ij]}} = G\ell^2 \eta_3 \kappa_{[ij]}, \quad \eta_3 > 0 \end{aligned} \tag{7.18}$$

where ℓ is a material constant with the dimension of length, called also *material or internal length*. Thus

$$w_3^{(el)} = \frac{1}{2} G\ell^2 \left(\kappa_{(mn)} \kappa_{(mn)} + \eta_2 \kappa_{(mm)} \kappa_{(nn)} + \eta_3 \kappa_{[ij]} \kappa_{[ij]} \right) \tag{7.19}$$

Note that the general anisotropic Cosserat elasticity can be found in a paper by Kessel [2].

7.2 A 2D Linear, Isotropic Cosserat-Elasticity Theory

Here we summarize some results from the paper of Schäfer [3], that pertain to a proposition for a simple two-dimensional, linear elasticity theory for isotropic materials of the Cosserat type. This is a theory of plane stress states.

First we assume that the symmetric part of the stress tensor is related to the symmetric part of the deformation tensor (i.e. to the symmetric part of the displacement gradient, that is identified with the infinitesimal strain tensor), through the usual equations of plane-stress, isotropic Hooke elasticity,

$$\begin{aligned} \varepsilon_{11} &= \frac{1}{E}(\sigma_{11} - \nu\sigma_{22}) \\ \varepsilon_{22} &= \frac{1}{E}(\sigma_{22} - \nu\sigma_{11}) \\ \varepsilon_{12} &= \frac{1}{2G}\sigma_{(12)} \end{aligned} \tag{7.20}$$

where

$$\varepsilon_{ij} = \frac{1}{2}\left(u_{i,j} + u_{j,i}\right) \tag{7.21}$$

The antisymmetric parts of the relative deformation and of the stress are also linked by a linear relation,

$$\gamma_{[12]} = \frac{1}{2G_c}\sigma_{[12]}, \quad G_c = \eta_1 G > 0 \tag{7.22}$$

where according to Eqs. (3.81) and (4.13)

$$\begin{aligned} \gamma_{[12]} &= \frac{1}{2}(\gamma_{12} - \gamma_{21}) = \omega - \psi \\ \sigma_{[12]} &= \frac{1}{2}(\sigma_{12} - \sigma_{21}) \end{aligned} \tag{7.23}$$

Due to the isotropy requirement we assume that the couple stresses are linked to the curvatures by means of only one additional material constant,

$$\mu_{13} = D\kappa_{13}; \quad \mu_{23} = D\kappa_{23}, \quad D > 0 \tag{7.24}$$

where

$$\begin{aligned} \kappa_{13} &= \partial_1\psi_3 = \partial_1\psi \\ \kappa_{23} &= \partial_2\psi_3 = \partial_2\psi \end{aligned} \tag{7.25}$$

7.3 Examples of Elementary Cosserat Elasticity B.V. Problems

7.3.1 Pure Bending of a Cosserat-Elastic Beam

We consider a beam with rectangular cross section, as shown in Fig. 7.1. The only stresses that are considered are the axial stress σ_{xx} and the couple stress μ_{xy}.

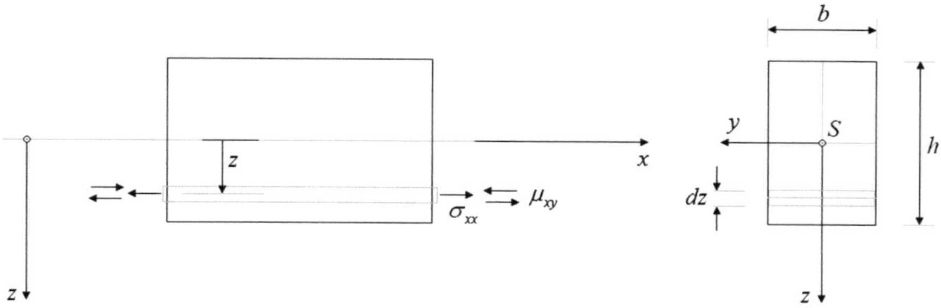

Fig. 7.1 Pure bending of a Cosserat beam: stresses on fiber

Motivated by the classical beam theory we apply a semi-inverse approach and assume that [3],

$$\sigma_{xx} = cz; \qquad \mu_{xy} = \tilde{c} \tag{7.26}$$

where c and $\tilde{c}$ are positive constants to be determined.

In the considered case the only significant equilibrium equations are,

$$\frac{\partial \sigma_{xx}}{\partial x} = 0; \quad \frac{\partial \mu_{xy}}{\partial x} = 0 \tag{7.27}$$

Thus the stress fields, Eqs. (7.26), are equilibrium fields.

The elasticity equations and the ansatz (7.26) yield,

$$\varepsilon_{xx} = \frac{1}{E}\sigma_{xx} = \frac{c}{E}z \tag{7.28}$$

$$\kappa_{xy} = \frac{1}{D}\mu_{xy} = \frac{\tilde{c}}{D} \tag{7.29}$$

The only surviving compatibility condition in the considered case is,

$$\overset{(2)}{I}_{21} = \frac{\partial \varepsilon_{xx}}{\partial z} - \kappa_{xy} = 0 \tag{7.30}$$

which in turn yields a restriction for the introduced constants,

$$\frac{c}{E} - \frac{\tilde{c}}{D} = 0 \quad \Rightarrow \quad \tilde{c} = \frac{D}{E}c \tag{7.31}$$

Note that in the paper by Schäfer [3] we find the derivation of stress functions that satisfy equilibrium and compatibility conditions.

We normalize the material constant D by the Young's modulus, by setting.

$$D = \frac{1}{12} E\ell^2 \tag{7.32}$$

where ℓ is a micro-mechanical length that is in most cases considered to be small, if compared with the typical geometric dimension of a structure. As we will see below, the factor $1/12$ is put for convenience in the computation.

With this remark from Eqs. (7.29) and (7.32) we get

$$\mu_{xy} = \frac{1}{12} E\ell^2 \kappa_{xy} \tag{7.33}$$

and

$$\kappa_{xy} = \frac{c}{E} \tag{7.34}$$

As in classical beam bending theory, from Fig. 7.2 we read

$$\begin{aligned} \frac{\partial_x u_x dx}{z} = \frac{dx}{R} \Rightarrow \frac{\varepsilon_{xx}}{z} = \frac{1}{R} \Rightarrow \\ \frac{c}{E} = \frac{1}{R} \end{aligned} \tag{7.35}$$

where R is the radius of curvature of the beam.

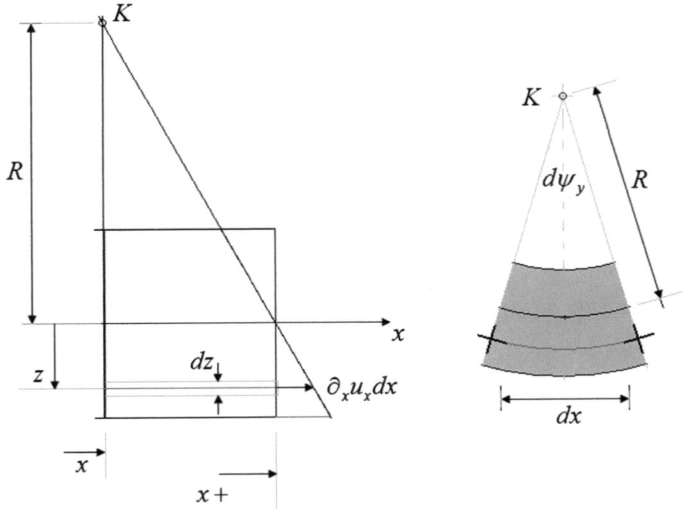

Fig. 7.2 Pure bending: exploitation of the Euler-Bernoulli hypothesis

On other hand, from Eqs. (7.35) and (7.34) we get

$$\frac{\partial \psi_y}{\partial x} = \frac{1}{R} \Rightarrow dx = R d\psi_y \tag{7.36}$$

The total bending moment that is taken by the rectangular cross-section of the beam is computed as

$$\begin{aligned} M &= \int_{-h/2}^{h/2} \sigma_{xx} z b dz + \int_{-h/2}^{h/2} \mu_{xy} b dz = cb \int_{-h/2}^{h/2} z^2 dz + \tilde{c} b \int_{-h/2}^{h/2} dz \\ &= c\frac{bh^3}{12} + \tilde{c} bh = c\left(\frac{bh^3}{12} + \frac{bh}{12}\ell^2\right) = cI\left(1 + \left(\frac{\ell}{h}\right)^2\right) \end{aligned} \tag{7.37}$$

where I is the surface moment of inertia of the rectangular cross-section of the beam,

$$I = \frac{bh^3}{12} \tag{7.38}$$

Then from Eq. (7.37) we get

$$M = \frac{EI'}{R} \tag{7.39}$$

where

$$I' = I\left(1 + \left(\frac{\ell}{h}\right)^2\right) \tag{7.40}$$

This is a typical result of Cosserat elasticity theory, meaning that a structure made of Cosserat elastic material is stiffer then the corresponding classical elastic structure. The smaller the structure is, the larger is the effect of the material length to the bending stiffness.

7.3.2 *Annular Shear of a Cylindrical Hole in Cosserat-Elastic Solid*

7.3.2.1 Background

The problem of annular shear of a cylindrical hole in Cosserat-elastic solids has been analyzed first by Besdo [6]. Bogdanova-Bontcheva and Lippmann [7], Unterreiner and Vardoulakis [8] analyzed the same problem within the frame of

Cosserat elastoplasticity. Controlled interface shear tests on granular materials were performed in the annular shear apparatus ACSA of CERMES/ENPC [4]; Fig. 7.3. When a granular material is sheared against a rough boundary, zones of localized deformation are observed at the interface; Fig. 7.4. The boundary localization phenomenon in the annular shear apparatus was simulated numerically by Zervos et al. [5] by using the Contact Dynamics method (Fig. 7.5). More recently Mohan et al. [9] have tackled this problem semi-analytically/numerically by utilizing the elasto-plastic model of Mühlhaus and Vardoulakis [10]. We quote from this paper [9]: *"...Experiments on viscometric flows of dense, slowly deforming granular materials indicate that shear is confined to a narrow region, usually a few grain diameters thick, while the remaining material is largely undeformed."*. In this section we address the problem analytically by using Cosserat elasticity in order to demonstrate that even the simplest Cosserat model will allow the formation of such "boundary layers".

Fig. 7.3 Plane-strain Couette-type annular shear apparatus for sand [4]

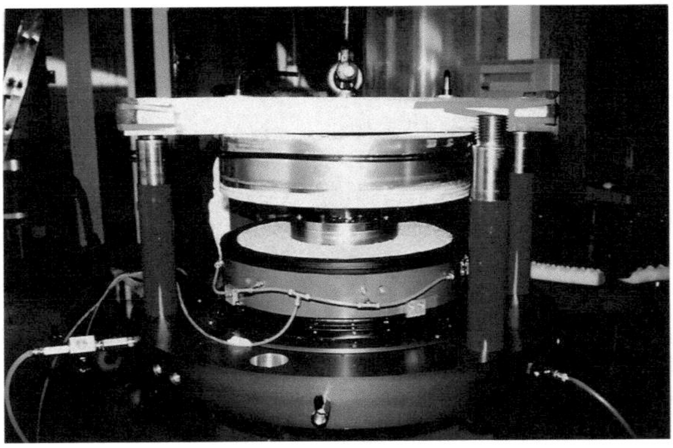

Fig. 7.4 Interface localization in granular material realized in the annular shear apparatus [4]

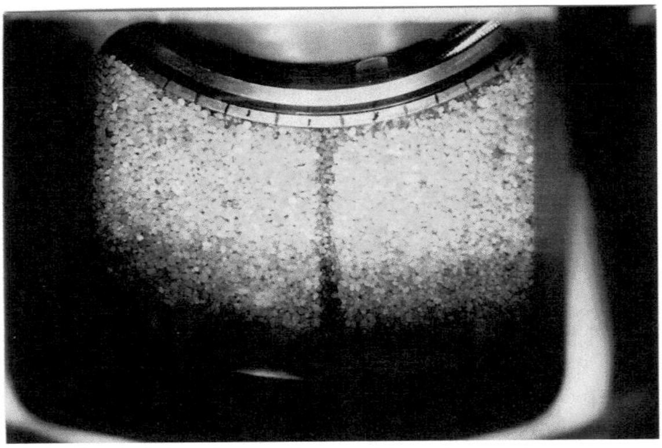

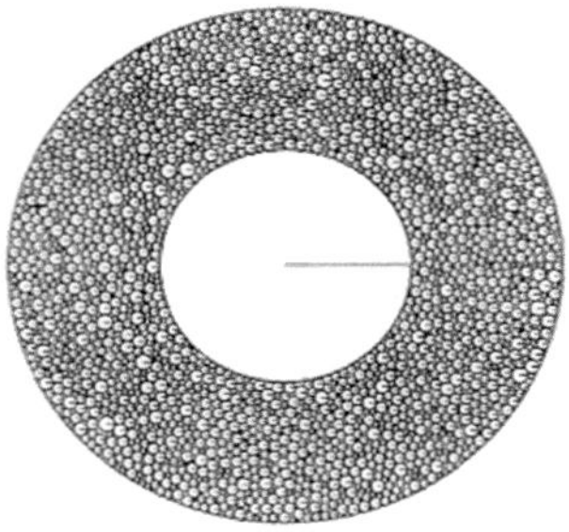

Initial State

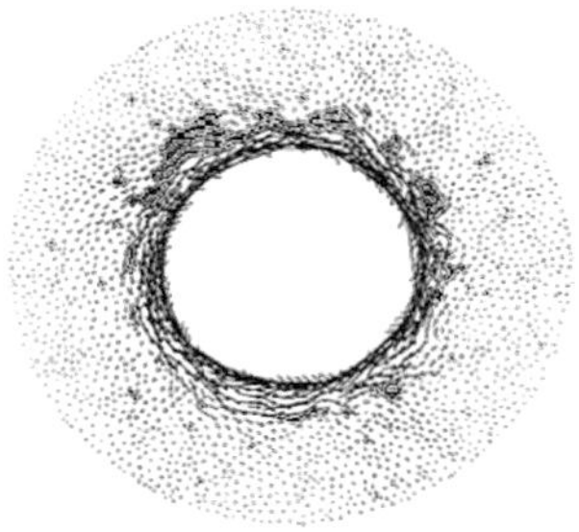

Total Displacements

Fig. 7.5 Contact dynamics simulation of interfacial localization [5]

7.3.2.2 Problem Statement

We consider a cylindrical hole of radius R under the action of internal shear as shown in Fig. 7.6. Axial symmetry yields the following equilibrium equations,

$$\frac{d\sigma_{r\theta}}{dr} + \frac{1}{r}(\sigma_{r\theta} + \sigma_{\theta r}) = 0 \tag{7.41}$$

$$\frac{d\mu_{rz}}{dr} + \frac{1}{r}\mu_{rz} + \sigma_{r\theta} - \sigma_{\theta r} = 0 \tag{7.42}$$

We note that for Boltzmann continua the considered problem is isostatic; i.e. for the determination of the stress field one does not need to specify the constitutive equation. Indeed in that case the only valid equilibrium equation is,

$$\frac{d\sigma_{r\theta}}{dr} + \frac{2}{r}\sigma_{r\theta} = 0 \tag{7.43}$$

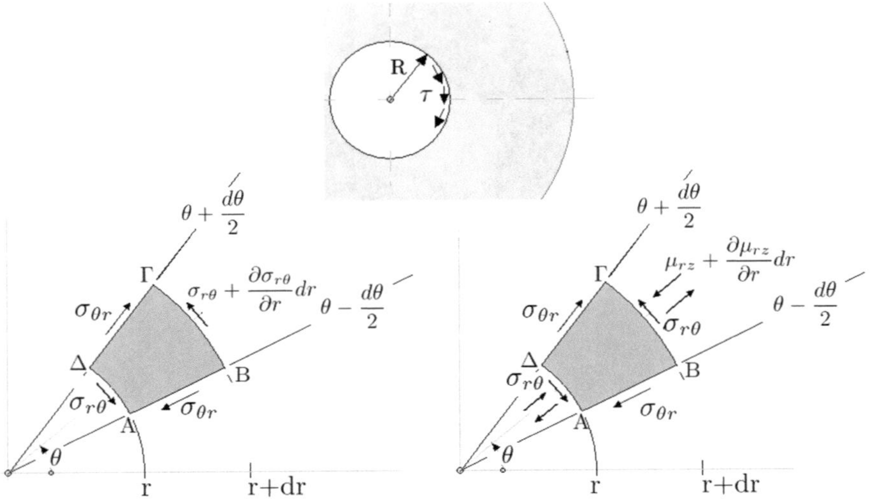

Fig. 7.6 Cylindrical hole in plane strain annular shear

The boundary conditions for the classical problem are,

$$r = R: \quad \sigma_{r\theta} = \tau \tag{7.44}$$

$$r \to \infty: \quad |\sigma_{r\theta}| < \infty \tag{7.45}$$

These b.c. admit the solution

$$\sigma_{r\theta} = \tau \left(\frac{r}{R}\right)^{-2} \tag{7.46}$$

As mentioned, the problem of a cylindrical cavity under annular shear is isostatic (Figs. 7.7 and 7.8). In this case principal stresses exist and their trajectories are logarithmic spirals,

$$\begin{aligned}
(\sigma^1): r &= \mathrm{R}\, exp\left(\theta \cot\left(\frac{\pi}{4}\right)\right), \quad x = r\cos(\theta + \theta_0), \quad y = r\sin(\theta + \theta_0) \\
(\sigma^2): r &= \mathrm{R}\, exp\left(-\theta \cot\left(\frac{3\pi}{4}\right)\right), \quad x = r\cos(-\theta + \theta_0), \quad y = r\sin(-\theta + \theta_0)
\end{aligned} \tag{7.47}$$

In the case, however, of a Cosserat continuum, none of the above holds. The problem is not isostatic, the solution depends on the constitutive assumptions made for the stresses and the couple stresses and a boundary layer is forming close to the cavity wall, where Cosserat effects are dominant. Indeed in case of a Cosserat continuum, we get the following expressions for the deformation measures,

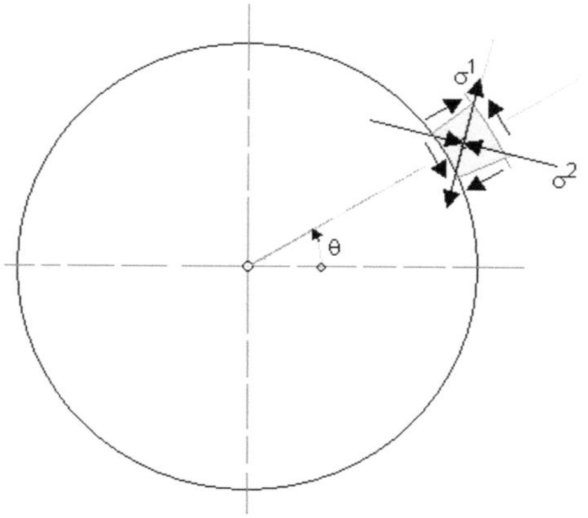

Fig. 7.7 Stress state at the cavity lips in case of a Boltzmann continuum

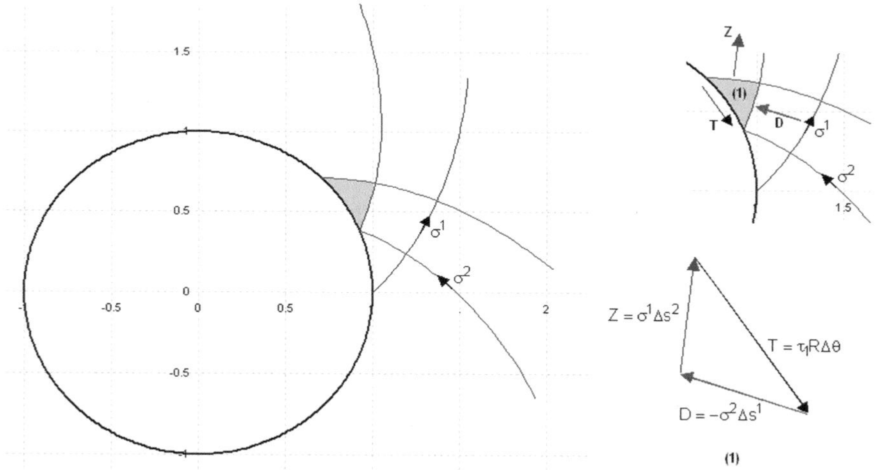

Fig. 7.8 Principal stress trajectories in case of Boltzmann continuum, indicating the isostaticity of the considered problem

$$sym[\gamma] = \begin{bmatrix} \varepsilon_{rr} & \varepsilon_{r\theta} & \varepsilon_{rz} \\ \varepsilon_{\theta r} & \varepsilon_{\theta\theta} & \varepsilon_{\theta z} \\ \varepsilon_{zr} & \varepsilon_{z\theta} & \varepsilon_{zz} \end{bmatrix} = \begin{bmatrix} 0 & \frac{1}{2}\left(\frac{\partial u_\theta}{\partial r} - \frac{u_\theta}{r}\right) & 0 \\ \frac{1}{2}\left(\frac{\partial u_\theta}{\partial r} - \frac{u_\theta}{r}\right) & 0 & 0 \\ 0 & 0 & 0 \end{bmatrix} \tag{7.48}$$

$$asym[\gamma] = \begin{bmatrix} 0 & \frac{1}{2}\left(\frac{\partial u_\theta}{\partial r} + \frac{u_\theta}{r}\right) - \psi_z & 0 \\ -\frac{1}{2}\left(\frac{\partial u_\theta}{\partial r} + \frac{u_\theta}{r}\right) + \psi_z & 0 & 0 \\ 0 & 0 & 0 \end{bmatrix} \tag{7.49}$$

$$[\kappa] = \begin{bmatrix} 0 & 0 & \frac{\partial \psi_z}{\partial r} \\ 0 & 0 & 0 \\ 0 & 0 & 0 \end{bmatrix} \tag{7.50}$$

These expressions are combined here with the constitutive equations of linear isotropic Cosserat elasticity and provide the following set of generalized stress-strain relationships,

$$\sigma_{(r\theta)} = 2G\varepsilon_{r\theta} = G\left(\frac{du_\theta}{dr} - \frac{u_\theta}{r}\right) \tag{7.51}$$

$$\sigma_{[r\theta]} = 2\eta_1 G\gamma_{[r\theta]} = \eta_1 G\left(\left(\frac{du_\theta}{dr} + \frac{u_\theta}{r}\right) - 2\psi_z\right) \tag{7.52}$$

and couple-stress-curvature relationships

$$\begin{aligned} \mu_{(rz)} &= G\ell^2\kappa_{(rz)} = \frac{1}{2}G\ell^2\frac{d\psi_z}{dr} \\ \mu_{[rz]} &= G\ell^2\eta_3\kappa_{[rz]} = \frac{1}{2}G\ell^2\eta_3\frac{d\psi_z}{dr} \end{aligned} \tag{7.53}$$

Thus

$$\sigma_{r\theta} = \sigma_{(r\theta)} + \sigma_{[r\theta]} = G\left((1+\eta_1)\frac{du_\theta}{dr} - (1-\eta_1)\frac{u_\theta}{r} - 2\eta_1\psi_z\right) \tag{7.54}$$

and

$$\mu_{rz} = \mu_{(rz)} + \mu_{[rz]} = \frac{1}{2}G\ell^2(1+\eta_3)\frac{d\psi_z}{dr} \tag{7.55}$$

Introducing the above set of constitutive equations into the equilibrium Eqs. (7.41) and (7.42) we get a set of two coupled differential equations for the particle displacement in tangential direction u_θ and for the particle rotation, ψ_z,

$$\frac{d^2u_\theta}{dr^2} + \frac{1}{r}\frac{du_\theta}{dr} - \frac{u_\theta}{r^2} = 2a\frac{d\psi_z}{dr} \tag{7.56}$$

and

$$\ell^2\frac{d^2\psi_z}{dr^2} + \ell^2\frac{1}{r}\frac{d\psi_z}{dr} - 2b\psi_z = -b\left(\frac{du_\theta}{dr} + \frac{u_\theta}{r}\right) \tag{7.57}$$

where

$$a = \frac{\eta_1}{1+\eta_1}; \quad b = 4\frac{\eta_1}{1+\eta_3} \tag{7.58}$$

For $\eta_1 = 0$ $(a = b = 0)$ the stress tensor is symmetric and Eqs. (7.56) and (7.57) become decoupled,

$$\frac{d^2 u_\theta}{dr^2} + \frac{1}{r}\frac{du_\theta}{dr} - \frac{u_\theta}{r^2} = 0 \tag{7.59}$$

$$\frac{d^2 \psi_z}{dr^2} + \frac{1}{r}\frac{d\psi_z}{dr} = 0 \tag{7.60}$$

The solution of Eqs. (7.59) and (7.60) is,

$$u_\theta = C_3 r + C_4 \frac{1}{r} \tag{7.61}$$

$$\psi_z = C_1 + C_2 \ln r \tag{7.62}$$

The boundedness condition at infinity for the particle rotation angle ψ_z and for the circumferential displacement u_θ is fulfilled, if $C_2 = 0$ and $C_3 = 0$. The solution for $C_1 \neq 0$ is physically meaningless, thus we accept the solution

$$u_\theta = \frac{C_4}{r}; \quad \psi_z = 0 \tag{7.63}$$

The integration constant C_4 is determined from the boundary condition for the shear stress,

$$r = R: \quad \sigma_{r\theta} = \tau \tag{7.64}$$

Thus

$$u_\theta = -\bar{u}\frac{R}{r}; \quad \bar{u} = \frac{\tau}{2G}R \tag{7.65}$$

In the uncoupled case $(\eta_1 = 0)$, the valid solution for the displacement, Eq. (7.65), together with the constitutive equation for the symmetric part of the stress, Eq. (7.51), yield the classical solution, Eq. (7.46).

In the general case $(\eta_1 > 0)$, Eqs. (7.57) and (7.56) yield

$$\ell^2\left(\frac{d^2\psi_z}{dr^2} + \frac{1}{r}\frac{d\psi_z}{dr}\right) - \eta^2\psi_z = 0 \tag{7.66}$$

where $\ell > 0$ and,

$$\eta^2 = 8\frac{\eta_1}{1+\eta_3}\frac{1}{1+\eta_1} > 0 \quad (\eta_1 > 0) \tag{7.67}$$

Let

$$\rho = \eta\frac{r}{\ell} \tag{7.68}$$

the general solution of Eq. (7.66) is given in terms of 0th-order modified Bessel functions

$$\psi_z = C_1 I_0(\rho) + C_2 K_0(\rho) \tag{7.69}$$

and from that

$$\frac{d\psi_z}{dz} = C_1\frac{\eta}{\ell}I_1(\rho) - C_2\frac{\eta}{\ell}K_1(\rho) \tag{7.70}$$

The extra boundary conditions are given in terms of the particle rotation and/or of the couple stress. In order to introduce these extra boundary conditions within the Cosserat continuum description, we resort to the concept of *ortho-fiber* [11]. An ortho-fiber is a rigid bar of length ℓ' aligned normally to the surface of the considered Cosserat continuum body and it is pointing outwards. On the end of this fiber we assume either displacements or tractions are applied thus giving to the surface actions the meaning of v. Mises motors (Fig. 7.9). Accordingly we assume here that at the cavity wall the shear stress and the couple stress are prescribed and at infinity the particle displacement and rotation must vanish,

$$r = R: \quad \begin{pmatrix} \sigma_{r\theta} \\ \mu_{rz} \end{pmatrix} = \begin{pmatrix} \tau \\ -\tau\ell' \end{pmatrix} \tag{7.71}$$

$$r = R^* \to \infty: \quad \begin{pmatrix} \psi_z \\ u_\theta \end{pmatrix} = \begin{pmatrix} 0 \\ 0 \end{pmatrix} \tag{7.72}$$

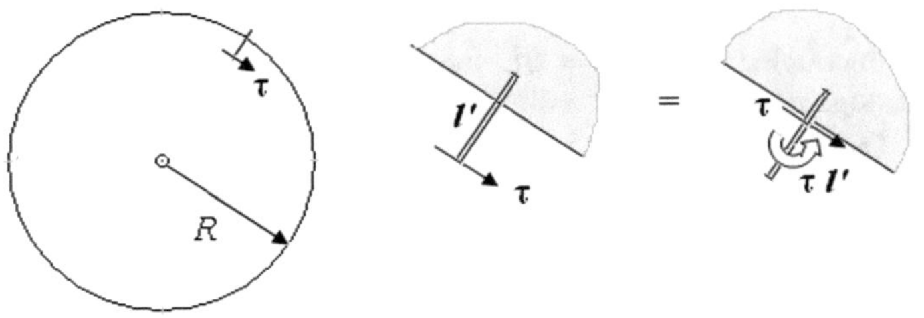

Fig. 7.9 Shear traction applied on an ortho-fiber at distance ℓ', resulting into a surface traction and a surface couple when transported to the surface

The sign of the surface couple in Eq. (7.71) results as follows: as shown in Fig. 7.9 the shear traction of magnitude τ is assumed to be applied on an ortho-fiber of length ℓ', thus yielding an equivalent set of surface actions, a surface traction and a surface couple. If the surface traction is positive then the surface couple must be negative.

For large argument ρ we have the following asymptotic expression for the solution, Eq. (7.69)

$$\psi_z = C_1 \frac{\exp(\rho)}{\sqrt{2\pi\rho}} + C_2 \sqrt{\frac{\pi}{2\rho}} \exp(-\rho) \tag{7.73}$$

From Eqs. (7.72), (7.71) and (7.73) we get

$$\begin{aligned} &e^{\rho^*} c_1 + e^{-\rho^*} c_2 = 0 \qquad (\rho^* \gg 1) \\ &I_1(\rho_R) c_1 + K_1(\rho_R) c_2 = -1 \end{aligned} \tag{7.74}$$

where

$$c_i = -\frac{C_i}{\frac{1}{(1+\eta_3)\eta} \frac{2\tau}{G} \frac{\ell'}{\ell}} \quad (i = 1, 2); \quad \rho_R = \eta \frac{R}{\ell} \tag{7.75}$$

The solution of the system of linear Eqs. (7.74) takes the following form,

$$\begin{aligned} c_1 &= \frac{e^{-2\rho^*}}{K_1(\rho_R) - I_1(\rho_R)e^{-2\rho^*}} \to 0 \quad (\rho^* \to \infty) \\ c_2 &= \frac{1}{K_1(\rho_R) - I_1(\rho_R)e^{-2\rho^*}} \to \frac{1}{K_1(\rho_R)} \quad (\rho^* \to \infty) \end{aligned} \tag{7.76}$$

This means that the valid solution here is the logarithmic one ($C_1 = 0$)

$$\psi_z = C_2 K_0(\rho) \tag{7.77}$$

and with that

$$\mu_{rz} = -C_2 \frac{\eta}{\ell} G \ell^2 (1 + \eta_3) K_1(\rho) \tag{7.78}$$

The b.c. at the cavity wall, Eq. (7.71) for the couple-stress yields,

$$C_2 = \frac{1}{\eta(1+\eta_3)} \left(\frac{4u}{R}\right) \frac{1}{K_1(\rho_R)} \left(\frac{\ell'}{\ell}\right) \tag{7.79}$$

and with that (Fig. 7.10)

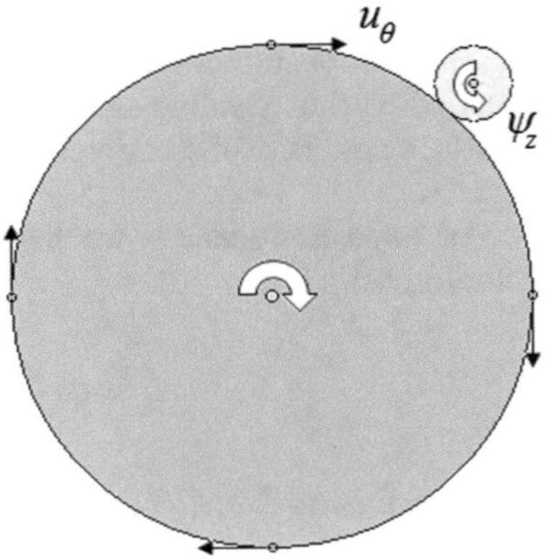

Fig. 7.10 Boundary particle displacement and rotation

$$\psi_z = \frac{1}{\eta(1+\eta_3)}\left(\frac{4u}{R}\right)\left(\frac{\ell'}{\ell}\right)\frac{K_0(\rho)}{K_1(\rho_R)} > 0 \tag{7.80}$$

For small values of the internal length $0 < \eta\ell \ll R$ we have the following asymptotic solution for the particle rotation.

$$\psi_z \approx 4\left(\frac{u}{R}\right)\left(\frac{\ell'}{\ell}\right)\sqrt{\frac{\rho_R}{\rho}}e^{-(\rho-\rho_R)} \tag{7.81}$$

This means that the particle rotations are confined in a boundary layer and they decay faster than exponentially with the radial distance from the cavity wall. On the other hand, we observe that the particle rotation depends linearly on the ratio of the "roughness" length scale ℓ' to the material length scale ℓ, and on the ratio of first-order imposed hoop displacement to the radius of the cavity.

Equation (7.56) becomes,

$$\frac{d^2u_\theta}{dr^2} + \frac{1}{r}\frac{du_\theta}{dr} - \frac{u_\theta}{r^2} - C\frac{d}{dr}K_0(\rho) = 0; \quad C = 2aC_2 \tag{7.82}$$

or

$$u_\theta = C_3 r + C_4\frac{1}{r} - 2aC_2\frac{\ell}{\eta}K_1(\rho) \tag{7.83}$$

As already explained above, the only meaningful solution is the one for $C_3 = 0$, thus

$$u_\theta = C_4\frac{1}{r} - 2\frac{C}{1+\eta_1}\frac{\ell}{\eta}\frac{K_1(\rho)}{K_1(\rho_R)} \tag{7.84}$$

where

$$C = C_2\eta_1 = \frac{4\eta_1}{\eta(1+\eta_3)}\frac{1}{K_1(\rho_R)}\frac{\bar{u}}{R}\frac{\ell'}{\ell} \tag{7.85}$$

The b.c. at the cavity wall, Eq. (7.71), for the shear stress yields,

$$C_4 = -u + CR^2\left(2\frac{1}{1+\eta_1}\frac{1}{\rho_R} + \frac{K_0(\rho_R)}{K_1(\rho_R)} - K_0(\rho_R)\right) \tag{7.86}$$

and with that

$$u_\theta \approx u_\theta^{(0)} - \tilde{u}_\theta \tag{7.87}$$

In this expression with $u_\theta^{(0)}$ we denote the classical solution Eq. (7.65)

$$u_\theta^{(0)} = -u\frac{R}{r}; \quad u = \frac{\tau}{2G}R \tag{7.88}$$

and $\tilde{u}_\theta$ is the perturbation that stems from the Cosserat terms,

$$\begin{aligned} \tilde{u}_\theta &\approx \frac{C\ell}{\eta}\left(\frac{\rho_R^2}{\rho} + \frac{2}{1+\eta_1}\left(\frac{\rho_R}{\rho} - \sqrt{\frac{\rho_R}{\rho}}e^{-(\rho-\rho_R)}\right)\right) \\ \frac{C\ell}{\eta} &= \frac{\eta_1}{\eta^2(1+\eta_3)}\frac{1}{K_1(\rho_R)}\frac{2\tau}{G}\ell' \end{aligned} \tag{7.89}$$

The perturbation for the displacement contains both hyperbolically and exponentially decaying terms, it is proportional to the classical solution, Eq. (7.65), and it scales linearly with the interfacial length scale, ℓ',

$$u_\theta = u_\theta^{(0)} - \tilde{u}_\theta \approx -\bar{u}\left(\frac{R}{r} + \tilde{C}\left(\frac{\ell'}{R}\right)f(\rho_R, \rho)\right) \tag{7.90}$$

7.3.3 *Sphere Under Uniform Radial Torsion*

We consider a spherical body of radius R made of linear-elastic, isotropic Cosserat-type material, that is subjected on its surface to uniform radial-torsional loading of intensity $\mu_{rr}(R) = m$ (Fig. 7.11). We want to analyze its state of stress and the deformation that this object suffers. In the considered setting the deformations, distortions and torsions are given in polar spherical coordinates as follows:

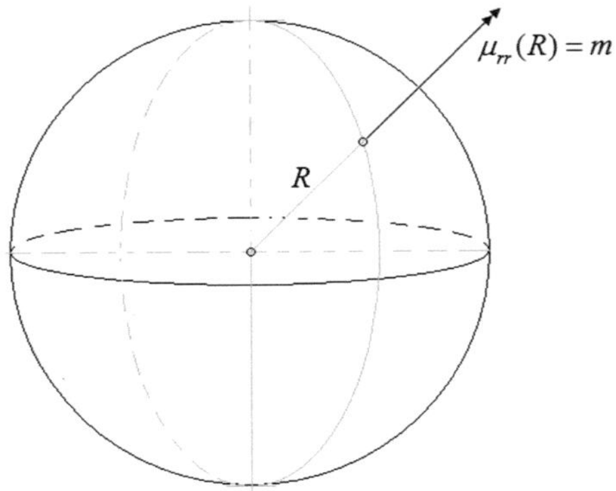

Fig. 7.11 Sphere under uniform surface torsion

$$asym[\gamma] = \begin{bmatrix} 0 & 0 & 0 \\ 0 & 0 & \psi_r \\ 0 & -\psi_r & 0 \end{bmatrix} \tag{7.91}$$

$$[\kappa] = \begin{bmatrix} \frac{d\psi_r}{dr} & 0 & 0 \\ 0 & \frac{\psi_r}{r} & 0 \\ 0 & 0 & \frac{\psi_r}{r} \end{bmatrix} \tag{7.92}$$

$$[\mu] = \begin{bmatrix} \mu_{rr} & 0 & 0 \\ 0 & \mu_{\theta\theta} & 0 \\ 0 & 0 & \mu_{\phi\phi} \end{bmatrix} \tag{7.93}$$

The only significant stress and couple-stress components are (Fig. 7.12):

$$\sigma_{\theta\phi} = \eta_1 G\psi_r; \quad \sigma_{\phi\theta} = -\eta_1 G\psi_r \tag{7.94}$$

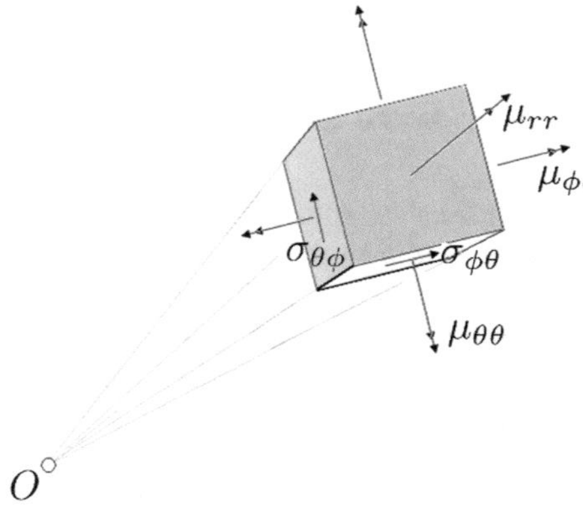

Fig. 7.12 Stress state in the element of a sphere under uniform torsion

$$\mu_{rr} = G\ell^2\left((1+\eta_2)\frac{d\psi_r}{dr} + 2\eta_2\frac{\psi_r}{r}\right)$$
$$\mu_{\theta\theta} = \mu_{\phi\phi} = G\ell^2\left(\eta_2\frac{d\psi_r}{dr} + (1+2\eta_2)\frac{\psi_r}{r}\right) \tag{7.95}$$

The governing equilibrium equation is:

$$\frac{d\mu_{rr}}{dr} + \frac{1}{r}(2\mu_{rr} - \mu_{\phi\phi} - \mu_{\theta\theta}) + \sigma_{\phi\theta} - \sigma_{\theta\phi} = 0 \tag{7.96}$$

The solution that is acceptable (as being regular at the origin) is (Fig. 7.13):

$$\psi_r = C\Psi(\rho); \quad \Psi = \frac{\rho\cosh\rho - \sinh\rho}{\rho^2}, \quad \rho = \frac{r}{\ell} \tag{7.97}$$

In this case the torsion is also confined in a boundary layer. For example the solution for the isotropic part of the torsion reads,

$$\mu_T = \frac{1}{3}(\mu_{rr} + \mu_{\theta\theta} + \mu_{\phi\phi}) = \frac{1}{3}G\ell^2(1+3\eta_2)\left(\frac{d\psi_r}{dr} + 2\frac{\psi_r}{r}\right) = C\mathrm{M}(\rho)$$
$$\mathrm{M} = \frac{\sinh\rho}{\rho} = +O(\rho^3) \tag{7.98}$$

Remark Following the seminal paper by Günther [12] a number of papers appeared, such as the ones by Grioli [13], Toupin [14], Mindlin and Tiersten [15], that dealt with the so-called *media with couple-stresses*. In this theory we assume

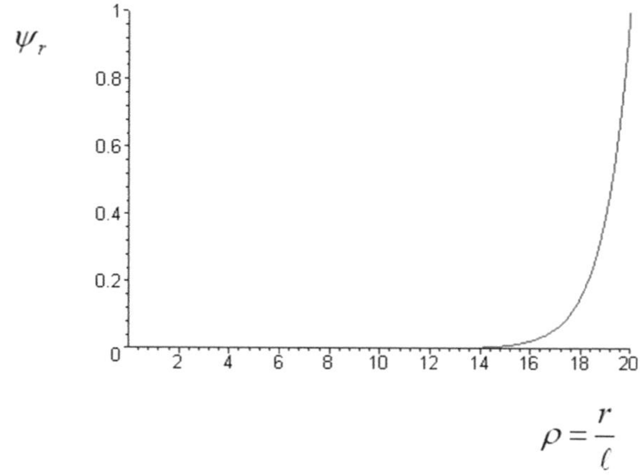

Fig. 7.13 Torsional boundary layer solution

that the relative particle spin is zero, meaning that the particle rotation coincides with the antisymmetric part of the displacement gradient. Such a theory is called also the *restricted Cosserat continuum or pseudo Cosserat continuum* [1] theory. In the original work of the Cosserat brothers, this special case is related to what they call the *triédre cache*. This theory, although widely used, has led to severe mathematical difficulties, since the isotropic part of the couple-stress remains indeterminate [16, 17] and cannot be determined by the boundary conditions. It is usually set arbitrarily equal to zero. This observation has lead to some controversies that we believe that have been resolved recently by Froiio et al. [11], who have shown, by using the concept of ortho-fiber, that such a restricted Cosserat continuum is incapable to absorb boundary conditions that refer to the torsional dof. In other words, the above example of the sphere under uniform torsion illustrates clearly the ability of the true Cosserat continuum theory to provide a unique solution for the torsion and the mean torsion!

References

1. Schaefer, H. (1967). Das Cosserat-Kontinuum. *Zeitschrift für Angewandte Mathematik und Mechanik, 47,* 485–498.
2. Kessel, S. (1964). Lineare Elastizitätstheorie des anisotropen Cosserat-Kontinuums. *Abhandlungen der Braunschweigische Wissenschaftliche Gesellschaft, 16,* 1–22.
3. Schaefer, H. (1962). Versuch einer Elastizitätstheorie des zweidimensionalen ebenen Cosserat-Kontinuums. In *Miszellaneen der angewandten Mechanik* (pp. 277–292). Berlin: Akademie-Verlag.
4. Corfdir, A., Lerat, P., & Vardoulakis, I. (2004). A cylinder shear apparatus. *ASTM Geotechnical Testing Journal, 27,* 1–9.
5. Zervos, A., et al. (2000). Numerical investigation of granular kinematics. *Mechanics of Cohesive-Frictional Materials, 5,* 305–324.
6. Besdo, D. (1974). Ein Beitrag zur nichtlinearen Theorie des Cosserat-Kontinuums. *Acta Mechanica, 20,* 105–131.
7. Bogdanova-Bontcheva, N., & Lippmann, H. (1975). Rotationssymmetrisches ebenes Fließen eines granularen Modellmaterials. *Acta Mechanica, 21,* 93–113.
8. Unterreiner, F., & Vardoulakis, I. (1994). Interfacial localisation in granular media. In *Proceedings of the eighth international conference on computer methods and advances in geomechanics*. Morgantown, West Virginia, USA: Balkema.
9. Mohan, S. L., Rao, K., & Nott, P. R. (2002). A frictional Cosserat model for the slow shearing of granular materials. *Journal of Fluid Mechanics, 457,* 377–408.
10. Mühlhaus, H.-B., & Vardoulakis, I. (1987). The thickness of shear bands in granular materials. *Géotechnique, 37,* 271–283.
11. Froiio, F., Zervos, A., & Vardoulakis, I. (2010). On natural boundary conditions in linear 2nd-grade elasticity. In *EUROMECH 510. Mechanics of generalized continua: A hundred years after the Cosserats*. Springer.
12. Günther, W. (1958). Zur Statik und Kinematik des Cosseratschen Kontinuums. *Abhandlungen der Braunschweigische Wissenschaftliche Gesellschaft, 10,* 195–213.
13. Grioli, G. (1960). Elasticità asimmetrica. *Anali di Matematica pura et applicata, 4,* 389–417.

14. Toupin, R. (1962). Elastic materials with couple-stresses. *Archive for Rational Mechanics and Analysis, 11,* 385–414.
15. Mindlin, R. D., & Tiersten, H. F. (1962). Effects of couple-stresses in linear elasticity. *Archive for Rational Mechanics and Analysis, 11,* 415–448.
16. Mindlin, R. D., & Eshel, N. N. (1968). On first strain-gradient theory in linear elasticity. *International Journal of Solids and Structures, 4,* 109–124.
17. Paria, G. (1970). Constitutive equations in Cosserat elasticity. *Journal of Engineering Mathematics, 4,* 203–208.

Chapter 8
Cosserat Fluids

Abstract This chapter moves from linear elastic solids to Cosserat fluids, showcasing the versatility of the approach. It presents the Navier-Stokes equations, generalized for an incompressible, linear viscous Cosserat fluid. Following this, examples are given, including shear flow and shallow flow slide of a granular fluid.

In the literature there is a number of publications concerning the Cosserat continuum generalization of the constitutive equations for simple fluids [1–3]. Here we consider the incompressible, linear Cosserat fluid.

8.1 Constitutive Equations

Let the Cartesian coordinates of the particle velocity and spin be $v_i(x_k, t)$ and $w_i(x_k, t)$ and let,

$$D_{ij} = \frac{1}{2}\left(\partial_i v_j + \partial_j v_i\right) \tag{8.1}$$

$$W_{ij} = \frac{1}{2}\left(\partial_j v_i - \partial_i v_j\right) \tag{8.2}$$

denote the classical rate-of-deformation and vorticity tensors, respectively. Note that the axial vector that is related to the spin tensor is computed as follows,

$$W_{ij} = -\varepsilon_{ijk}\dot{\omega}_k \quad \Leftrightarrow \quad \dot{\omega}_i = \frac{1}{2}\varepsilon_{ikl}\partial_k v_l \tag{8.3}$$

The rates of relative deformation- and distortion tensors are,

$$\Gamma_{ik} = \partial_i v_k - \varepsilon_{ikl} w_l \tag{8.4}$$

$$\mathrm{K}_{ij} = \partial_i w_j \tag{8.5}$$

I. Vardoulakis, *Cosserat Continuum Mechanics*, Lecture Notes in Applied and Computational Mechanics 87, https://doi.org/10.1007/978-3-319-95156-0_8

We decompose the rate of deformation-and the rate of distortion tensors into spherical and deviatoric part,

$$D_{ij} = \frac{1}{3} D_{kk} \delta_{ij} + D'_{ij} \quad ; \quad \mathrm{K}_{ij} = \frac{1}{3} \mathrm{K}_{kk} \delta_{ij} + \mathrm{K}'_{ij} \tag{8.6}$$

The constitutive equations for the stress and couple-stress for a linear, incompressible Cosserat fluid are,

$$\sigma_{ij} = -p\delta_{ij} + 2\mu D'_{ij} + 2\mu_c \varepsilon_{ijk}(\dot{\omega}_k - w_k) \tag{8.7}$$

$$\mu_{ij} = c_0 \mathrm{K}_{kk} \delta_{ij} + c_1 \mathrm{K}'_{(ij)} + c_2 \mathrm{K}'_{[ji]} \tag{8.8}$$

Due to the incompressibility constraint,

$$D_{kk} = 0 \tag{8.9}$$

the fluid pressure p is kinematically undetermined; it is determined by the boundary conditions of any given problem. For the same reason the rate-of-deformation tensor collapses to its deviator

$$D_{ij} = D'_{ij} \tag{8.10}$$

The constitutive Eq. (8.7) can be written also in the following form,

$$\begin{aligned} \sigma_{ij} &= -p\delta_{ij} + 2\mu D_{ij} - 2\mu_c \left(W_{ij} + \varepsilon_{ijk} w_k \right) \\ &= -p\delta_{ij} + 2\mu \Gamma_{(ij)} + 2\mu_c \Gamma_{[ij]} \\ &= -p\delta_{ij} + (\mu + \mu_c)\Gamma_{ij} + (\mu - \mu_c)\Gamma_{ji} \end{aligned} \tag{8.11}$$

If the particle spins as its neighbourhood, then Eqs. (8.11) collapse to those of a classical (Boltzmann) incompressible Newtonian fluid. Thus, in Eqs. (8.7) or (8.11), the constitutive parameter $\mu > 0$ is identified as the classical (macroscopic) fluid viscosity,

$$\sigma_{(ij)} = 2\mu D'_{ij} \tag{8.12}$$

The viscosity parameter μ_c is an extra material parameter that accounts for the relative spin of the Cosserat fluid particle with respect to background vorticity that is due to the particle velocity,

$$\sigma_{[ij]} = 2\mu_c \varepsilon_{ijk}(\dot{\omega}_k - w_k) \tag{8.13}$$

Since there is an axial vector assigned to the antisymmetric part of the stress tensor, Eq. (4.16), the extra constitutive Eq. (8.13) is a linear vector relation,

$$t_k^* = 2\mu_c(\dot{\omega}_k - w_k) \tag{8.14}$$

We can normalize this extra viscosity by the macroscopic viscosity and write,

$$\mu_c = \alpha\mu \quad (\alpha > 0) \tag{8.15}$$

The constitutive equations for the couple-stress tensor can be also written as follows,

$$\begin{aligned}\mu_{ij} &= \left(c_0 - \frac{1}{3}c_1\right)\mathrm{K}_{kk}\delta_{ij} + \frac{1}{2}(c_1 + c_2)\mathrm{K}_{ij} + + \frac{1}{2}(c_1 - c_2)\mathrm{K}_{ij} \\ &= c_T\partial_k w_k\delta_{ij} + c_B\partial_i w_j + c_B'\partial_j w_i\end{aligned} \tag{8.16}$$

They introduce three additional *gyro-viscosity* coefficients, which are taken proportional to the square of a material length parameter ℓ,

$$c_T = \gamma\mu\ell^2 \quad ; \quad c_B = \beta\mu\ell^2 \quad ; \quad c_{B'} = \beta'\mu\ell^2 \quad (\gamma, \beta, \beta' > 0) \tag{8.17}$$

Note that the above constitutive equations are proposed for the description of granular flow problems, where the material length is set equal to the (mean) grain diameter [4].

In summary, if we choose as primary kinematic variables the rate of relative deformation and the rate of distortion, then the constitutive equations for an incompressible, linear Cosserat fluid are the following,

$$\begin{aligned}\sigma_{ij} &= -p\delta_{ij} + (\mu + \mu_c)\Gamma_{ij} + (\mu - \mu_c)\Gamma_{ji} \\ \mu_{ij} &= c_T\mathrm{K}_{kk}\delta_{ij} + c_B\mathrm{K}_{ij} + c_B'\mathrm{K}_{ji} \\ \Gamma_{kk} &= 0\end{aligned} \tag{8.18}$$

8.2 Cosserat Generalization of the N.-S. Equations

We consider the dynamic Eq. (5.26) for the stresses,

$$\rho\frac{Dv_k}{Dt} = \partial_i\sigma_{ik} + \rho g_k \tag{8.19}$$

where ρ is the fluid density and g_k is the gravity acceleration. We introduce into the dynamic Eq. (8.19) the constitutive Eq. (8.7) and utilize Eqs. (8.1)–(8.10), thus yielding,

$$\rho \frac{Dv_k}{Dt} = -\partial_k p + \partial_i(\mu(\partial_i v_k + \partial_k v_i)) + \partial_i(\mu_c((\partial_i v_k - \partial_k v_i) - 2\varepsilon_{ikl} w_l)) + \rho g_k \quad (8.20)$$

For constant viscosities, Eq. (8.20) becomes,

$$\rho \frac{Dv_k}{Dt} = -\partial_k p + \mu(\partial_i \partial_i v_k + \partial_i \partial_k v_i) + \mu_c((\partial_i \partial_i v_k - \partial_i \partial_k v_i) - 2\varepsilon_{ikl} \partial_i w_l) + \rho g_k \quad (8.21)$$

From the incompressibility condition, Eq. (8.9), we get

$$D_{kk} = \partial_k v_k = 0 \quad \Rightarrow \quad \partial_i \partial_k v_i = \partial_k \partial_i v_i = 0 \quad (8.22)$$

Thus from Eqs. (8.21) and (8.22) we obtain,

$$\rho \frac{Dv_i}{Dt} + \partial_i p - \rho g_i = (\mu + \mu_c) \nabla^2 v_i + 2\mu_c \varepsilon_{ikl} \partial_k w_l \quad (8.23)$$

In Chapman-Cowling bold letter notation Eq. (8.23) reads as follows,

$$\rho \frac{D\boldsymbol{v}}{Dt} + \mathbf{grad} p - \rho \boldsymbol{g} = (\mu + \mu_c) \nabla^2 \boldsymbol{v} + 2\mu_c \mathbf{rot}\, \boldsymbol{w} \quad (8.24)$$

Equations (8.23) or (8.24) can be viewed as the Cosserat continuum generalization of the Navier-Stokes equations. Indeed for $\mu_c \to 0$ they reduce to the N.-S. equations for an incompressible Newtonian fluid

$$\rho \frac{Dv_i}{Dt} + \partial_i p - \rho g_i = \mu \nabla^2 v_i \quad (8.25)$$

We recall that the N.-S. equations are already a singular perturbation of the Euler equations for an ideal, incompressible fluid, that can be derived from Eq. (8.25) by setting $\nu = 0$,

$$\rho \frac{Dv_i}{Dt} + \partial_i p - \rho g_i = 0 \quad (8.26)$$

If we apply the rot-operator on both sides of Eq. (8.23) we get,

$$\rho \frac{D\dot{\omega}_i}{Dt} = (\mu + \mu_c) \nabla^2 \dot{\omega}_i + \mu_c \left(\nabla^2 w_i - \partial_i \partial_k w_k\right) \quad (8.27)$$

In bold notation Eq. (8.27) takes the form,

$$\rho \frac{D\dot{\boldsymbol{\omega}}}{Dt} = (\mu + \mu_c) \nabla^2 \dot{\boldsymbol{\omega}} + \mu_c \left(\nabla^2 \boldsymbol{w} - \mathbf{grad}\, \mathrm{div}\, \boldsymbol{w}\right) \quad (8.28)$$

This is a generalized vorticity diffusion equation with a source/sink term that is due to the particle spin. Note that, as already stated in Sect. 3.2.1, the last term on the r.h.s. of Eq. (8.27), can be interpreted as the gradient of the mean torsion. It is obvious that this term is in general non-zero. For $\mu_c = 0$ Eq. (8.27) or (8.28) reduce to the vorticity diffusion equation of classical Fluid Dynamics,

$$\rho \frac{D\dot{\omega}_i}{Dt} = \mu \nabla^2 \dot{\omega}_i \tag{8.29}$$

The second equation is derived from the dynamic Eqs. (5.70) for the couple-stresses in the absence of body couples, for constant viscosities and for isotropic micro-inertia tensor,

$$\rho J \frac{Dw_i}{Dt} = \partial_k \mu_{ki} + \varepsilon_{ikl} \sigma_{kl} \tag{8.30}$$

Introducing in this equation the constitutive equations for the couple-stresses, Eq. (8.16), and for the antisymmetric part of the stress, Eq. (8.13), we obtain the following,

$$\rho J \frac{Dw_k}{Dt} = c_B \nabla^2 w_k + (c_T + c_{B'}) \partial_k \partial_l w_l + 4\mu_c (\dot{\omega}_k - w_k) \tag{8.31}$$

or symbolically,

$$\rho J \frac{D\boldsymbol{w}}{Dt} = c_B \nabla^2 \boldsymbol{w} + \left(c_T + c_B'\right) \mathbf{grad}\, \mathrm{div}\, \boldsymbol{w} + 2\mu_c \mathbf{rot}\, \boldsymbol{v} - 4\mu_c \boldsymbol{w} \tag{8.32}$$

Equation (8.31) or (8.32) is the diffusion equation for the particle vorticity with a source/sink term that expresses the weak coupling to the generalized N.-S. Equations (8.27) or (8.28). From Eqs. (8.27) and (8.31) we observe that for vanishing macroscopic viscosity ($\mu \to 0$), not only the classical spin but also the Cosserat spin are sustained for "long" time intervals.

8.3 Shear Flow of a Cosserat Fluid

8.3.1 Kinematics and Statics of Forming Boundary Layers

As shown in Fig. 8.1, we consider a "long" layer of Cosserat fluid, confined between two "rough"plates, at distance H apart. Through appropriate boundary conditions, to be specified below, we assume that a steady, laminar shear flow is established. The only two significant kinematic variables of this problem are the

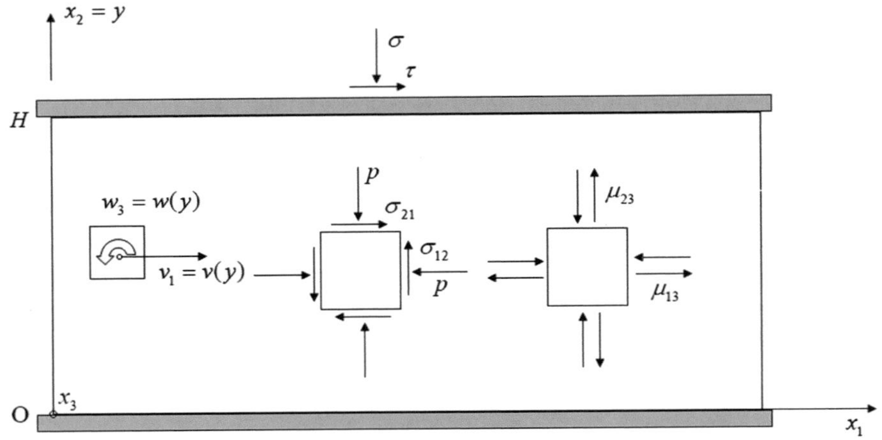

Fig. 8.1 A Cosserat-fluid shear layer

particle velocity in the long, x_1- direction and the particle spin in the x_3- direction, perpendicular to the plane of shear. Both fields are assumed to be functions only of $x_2 = y$,

$$v_1 = v(y) \tag{8.33}$$

$$w_3 = w(y) \tag{8.34}$$

From Eqs. (8.1), (8.2), (8.33) and (8.34) we get the following expression for the rate-of-deformation tensor,

$$[D] = [D'] = \begin{bmatrix} 0 & \frac{1}{2}v' & 0 \\ \frac{1}{2}v' & 0 & 0 \\ 0 & 0 & 0 \end{bmatrix} \tag{8.35}$$

where

$$(.)' \equiv \frac{d}{dy} \tag{8.36}$$

The flow is isochoric and the only significant component of the classical vorticity tensor is $\dot{\omega}_{21} = -\dot{\omega}_{12} = \dot{\omega}$,

$$[W] = \begin{bmatrix} 0 & \frac{1}{2}v' & 0 \\ -\frac{1}{2}v' & 0 & 0 \\ 0 & 0 & 0 \end{bmatrix} = \begin{bmatrix} 0 & -\dot{\omega} & 0 \\ \dot{\omega} & 0 & 0 \\ 0 & 0 & 0 \end{bmatrix} \tag{8.37}$$

Similarly, from Eqs. (8.4) and (8.5) we get the following matrix representations of the rate of relative deformation-and distortion tensors

$$[\Gamma] = \begin{bmatrix} 0 & -w & 0 \\ v' + w & 0 & 0 \\ 0 & 0 & 0 \end{bmatrix} \tag{8.38}$$

$$[\mathrm{K}] = \begin{bmatrix} 0 & 0 & 0 \\ 0 & 0 & w' \\ 0 & 0 & 0 \end{bmatrix} \tag{8.39}$$

From Eqs. (8.18), (8.38) and (8.39) we obtain the expressions for the non-zero stress and couple-stress components,

$$\sigma_{11} = \sigma_{22} = \sigma_{33} = -p \tag{8.40}$$

$$\sigma_{12} = (\mu - \mu_c)v' - 2\mu_c w \tag{8.41}$$

$$\sigma_{21} = (\mu + \mu_c)v' + 2\mu_c w \tag{8.42}$$

and

$$\mu_{23} = c_B w' \tag{8.43}$$

Thus the symmeric and antisymmetric parts of the stress are,

$$\sigma_{(12)} = \mu v' \tag{8.44}$$

$$\sigma_{[12]} = -\mu_c(v' + 2w) \tag{8.45}$$

Here gravity forces are either considered as being negligible or they are acting in the x_3- direction, normal to the plane of deformation, as is the case in an Couette-Hatschek type of apparatus (cf. Fig. 7.3). For shear-flow and in the absence of gravity forces and body couples, the equilibrium Eqs. (4.42) read,

$$\frac{\partial \sigma_{21}}{\partial y} = 0 \quad \Rightarrow \quad \sigma_{21} = \tau = const. \tag{8.46}$$

$$-\frac{\partial p}{\partial y} = 0 \quad \Rightarrow \quad p = const. \tag{8.47}$$

$$\frac{\partial \mu_{23}}{\partial y} + \sigma_{12} - \sigma_{21} = 0 \tag{8.48}$$

The pressure p is determined from the boundary condition for the normal stress that is applied on the upper plate,

$$\sigma_{22}(H) = -\sigma \quad (\sigma > 0) \quad \Rightarrow \quad p = \sigma = const. \tag{8.49}$$

Equations (8.46) and (8.42) yield,

$$v' = \frac{\tau}{\mu + \mu_c} - 2\frac{\mu_c}{\mu + \mu_c} w \tag{8.50}$$

Similarly for constant gyro-viscosity from Eqs. (8.48) and (8.43) we get

$$c_B w'' - 2\mu_c(v' + 2w) = 0 \tag{8.51}$$

The governing equation is derived by elimination of v' from Eqs. (8.50) and (8.51), resulting to,

$$c_B w'' - 4\frac{\mu_c}{\mu + \mu_c}\mu w = 2\frac{\mu_c}{\mu + \mu_c}\tau \tag{8.52}$$

If we scale the $y-$ coordinate with the geometric length that characterizes the boundary-value problem at hand; i.e. with layer thickness H,

$$\tilde{y} = \sqrt{2}\frac{y}{H} \tag{8.53}$$

Equation (8.52) becomes,

$$Co\frac{d^2 w}{d\tilde{y}^2} w'' - 2\frac{\alpha}{1+\alpha} w = \frac{\alpha}{1+\alpha}\frac{\tau}{\mu} \tag{8.54}$$

The parameter α is the viscosity ratio, that was introduced already above by Eq. (8.15). The dimensionless group,

$$Co = \frac{c_B}{\mu H^2} \tag{8.55}$$

is called the *Cosserat number* of the flow [3]. We note that with Eq. (8.17) the Cosserat number is in fact the square of a non-dimensional material length of the Cosserat fluid. The reference dimension of the domain H is assumed to be sufficiently "large", if compared to this micromechanical length parameter, and with that Co is a small number.

Equation (8.54) belongs to the set that is known to describe the formation of two boundary layers at $y = 0$ and $y = H$ [5]. Accordingly we use here terminology that considers the two asymptotic solutions of this equation, one, called the *outer approximation* and holding away from the boundaries and another called the *inner approximation* and holding close to the boundaries.

With $Co \to 0$ we get from Eq. (8.54) the outer approximation,

$$w \approx w^{(out)} = -\frac{1}{2}\frac{\tau}{\mu} \tag{8.56}$$

From Eqs. (8.50) and (8.56) we recover Newton's law,

$$\frac{dv}{dy} \approx \frac{\tau}{\mu} \tag{8.57}$$

In classical theory from Eq. (8.57) and Stoke's non-slip boundary condition we obtain the classical linear profile, as the corresponding outer approximation for the particle velocity,

$$v \approx v^{(out)} = \frac{\tau}{\mu} y \tag{8.58}$$

Note that,

$$\dot{\omega}^{(out)} = -\frac{1}{2}\frac{dv^{(out)}}{dy} = -\frac{1}{2}\frac{\tau}{\mu} = const. \tag{8.59}$$

is the only significant component of the background vorticity in the classical fluid. Thus the outer approximation for the particle spin, Eq. (8.56) takes the form,

$$w^{(out)} = \dot{\omega}^{(out)} = -\frac{1}{2} v^{(out)} \tag{8.60}$$

This means that away from the boundaries of the shear layer the fluid particles rotate as their neighbourhood.

The exact solution is obtained by the scaling of the $y-$ coordinate with the material dimension

$$y^* = \sqrt{2}\frac{y}{\ell_c} \tag{8.61}$$

where

$$\ell_c^2 = \frac{1}{2}\frac{\mu + \mu_c}{\mu_c} c_B = \frac{1}{2}\frac{1+\alpha}{\alpha}\beta\ell^2 \tag{8.62}$$

With this transformation Eq. (8.52) becomes,

$$\frac{d^2 w}{dy^{*2}} - w + w^{(0)} = 0 \tag{8.63}$$

The exact, general solution of this equation is,

$$w = w^{(out)} + C_1 \sinh y^* + C_1 \cosh y^* \tag{8.64}$$

The integration constants C_1 and C_2 must be determined from boundary conditions that refer to the spin. We note first that Stoke's classical non-slip condition applies to the particle velocity, and is expressed by taking all components of the particle velocity at a boundary equal to zero. As we saw clearly in Sect. 2.3.2, the velocity and the spin of a rigid particle are coupled through the well known transport law of rigid-body dynamics. Thus, one important implication of Stoke's non-slip condition is that, within the realm of Cosserat Continuum Fluid Mechanics, we must equally apply the corresponding non-spin boundary condition. Actually the non-spin boundary condition will imply the non-slip condition for the particle that adheres to a *rough* boundary. This condition is called in the literature *hyper-stick or adherence condition.*

In the considered boundary value problem the particle non-spin boundary conditions are,

$$w(0) = 0 \wedge w(H) = 0 \tag{8.65}$$

These boundary conditions are yielding the following exact solution (Fig. 8.2),

$$w = -\frac{1}{2}\frac{\tau}{\mu}(1 - f^*) \quad ; \quad f^* = \frac{\sinh y^* + \sinh(H^* - y^*)}{\sinh H^*} \tag{8.66}$$

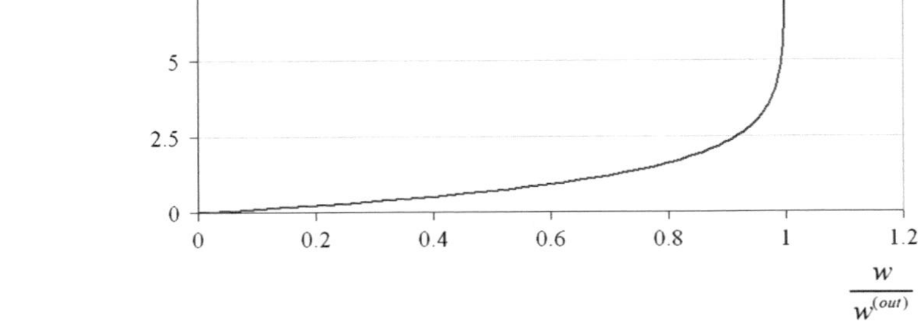

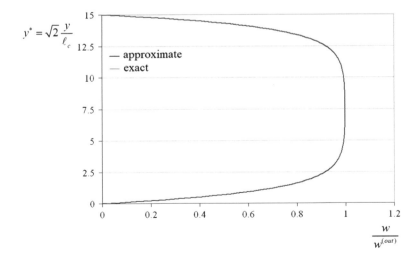

Fig. 8.2 Particle spin distribution for $\alpha = 1.$; Eqs. (8.66) and (8.71)

For relatively large values of the dimensionless layer thickness,

$$H^* = \sqrt{2}\frac{H}{\ell_c} \tag{8.67}$$

we get from Eq. (8.66) that close to the boundary at $y = 0$ the exact solution is approximated as,

$$\begin{aligned} f^* &= \frac{\sinh y^* + \sinh H^* \cosh y^* - \sinh y^* \cosh H^*}{\sinh H^*} \\ &\approx \cosh y^* - \sinh y^* = e^{-y^*} = f_0^* \end{aligned} \tag{8.68}$$

Similarly close to the boundary $y = H$ we get,

$$\begin{aligned} f^* &= \frac{\sinh y^* + \sinh H^* \cosh y^* - \sinh y^* \cosh H^*}{\sinh H^*} \\ &\approx \frac{\sinh y^*}{\sinh H^*} = \frac{\frac{1}{2}\left(e^{y^*} + e^{-y^*}\right)}{\frac{1}{2}\left(e^{H^*} + e^{-H^*}\right)} \approx e^{-(H^*-y^*)} = f_H^* \end{aligned} \tag{8.69}$$

From the above considerations we obtain the following asymptotic form for the considered function

$$f^* \approx f_0^* + f_H^* = e^{-y^*} + e^{-(H^*-y^*)} \quad , \quad 0 \le y^* \le H^* \tag{8.70}$$

and with that,

$$w \approx w^{(out)}\left(1 - \left(e^{-y^*} + e^{-(H^*-y^*)}\right)\right) \tag{8.71}$$

The difference between the two asymptotic solutions, $\left|w^{(out)} - w\right|$, is less then 1% at a distance from the boundaries,

$$d_{bl} \approx 4.6\frac{\ell_J}{\sqrt{2}} \approx 3.2\ell_c \tag{8.72}$$

that determines in turn the conventional thickness of the boundary layers. The material length parameter ℓ_c is set proportional to the grain size. In Fig. 8.3 we plotted this result as a function of the viscosity parameter α and for two typical values of the viscosity parameter β. Experimental observations suggest that a shear boundary layer in granular media has about half the thickness of the "shear-band", which in turn is about 10 to 15 grain diameters thick [6]. This gives rise to suggest that the parameter β should be of the same order as the parameter α.

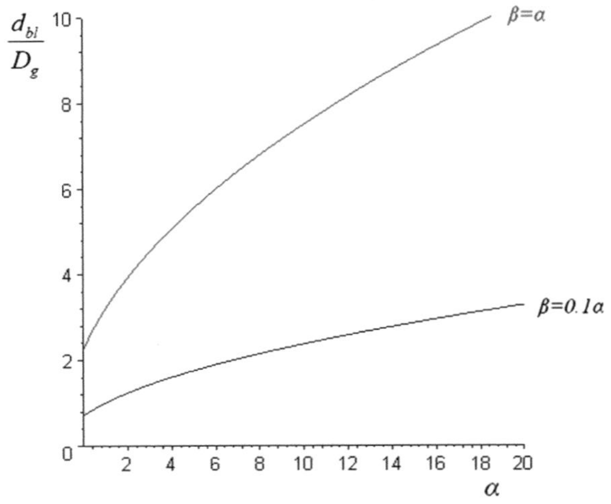

Fig. 8.3 Estimate of the boundary layer "thickness"

From Eqs. (8.50) and (8.66) we get,

$$v' = \frac{\tau}{\mu}\left(1 + \frac{\alpha}{1+\alpha}f^*\right) \tag{8.73}$$

Similarly from Eq. (8.73) we get that the shear rate v' is practically constant away from the boundaries (Fig. 8.4),

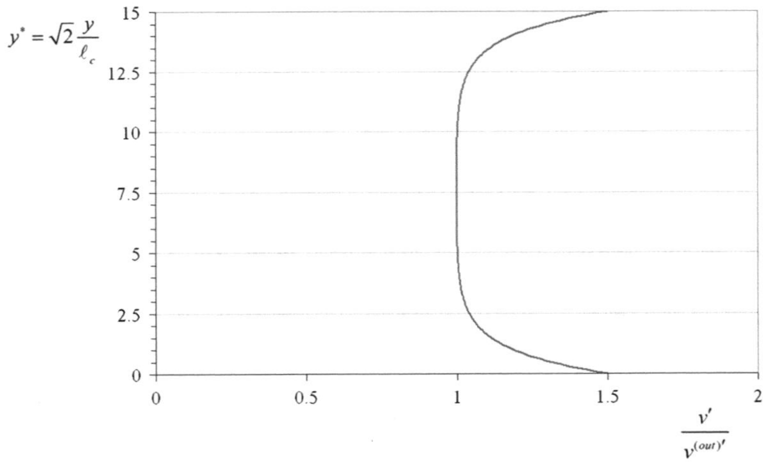

Fig. 8.4 Shear rate profile for $\alpha = 1.$; Eq. (8.73)

$$v' \approx v'^{(out)} = \frac{v^{(out)}}{H} = \frac{\tau}{\mu} \tag{8.74}$$

Close to the boundaries the shear rate assumes a value that is higher than $v'^{(out)}$ in proportion to the Cosserat viscosity ratio α; e.g. close to the boundary $y = 0$ we have that

$$v' \approx v'^{(out)}\left(1 + \frac{\alpha}{1+\alpha}e^{-y^*}\right) \approx v'^{(out)}\left(\frac{1+2\alpha}{1+\alpha} - O(y^*)\right) \tag{8.75}$$

As is shown in Fig. 8.5, the velocity profile is linear in the core of the shear-layer and convex inside the boundary layers (Fig. 8.5),

$$v \approx \frac{\tau}{\mu}\frac{\ell_c}{\sqrt{2}}\left(y^* + \frac{\alpha}{1+\alpha}\left(\left(1 - e^{-y^*}\right) - \left(e^{-H^*} - e^{-(H^*-y^*)}\right)\right)\right) \tag{8.76}$$

Finally, from Eqs. (8.43) and (8.66) we get also the corresponding expression for the couple stresses that are acting on planes parallel to the shear-layer axis (Fig. 8.6),

$$\mu_{23} = \beta\mu\ell^2 w' = \frac{1}{2}\tau\beta\ell^2 f^{*\prime} \approx -\frac{1}{2}\tau\beta\ell^2\left(e^{-y^*} - e^{-(H^*-y^*)}\right) \tag{8.77}$$

Thus couple stresses exist practically inside the two boundary layers.

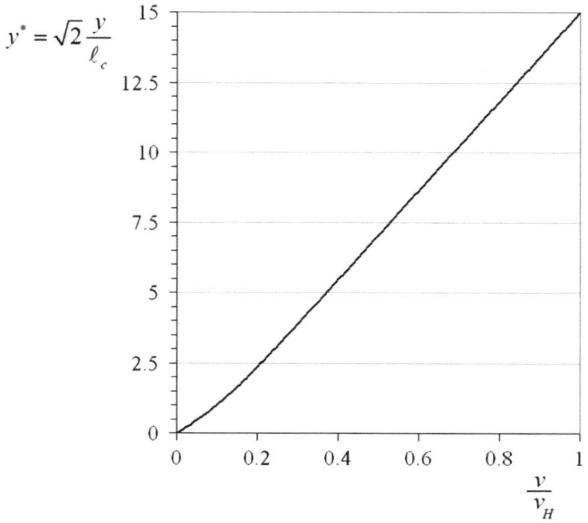

Fig. 8.5 Typical, locally convex/linear velocity profile, computed for $\alpha = 1.$; Eq. (8.76)

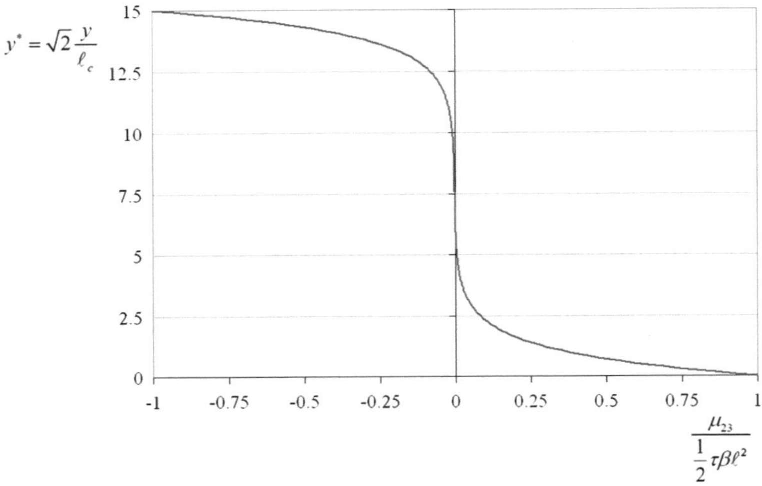

Fig. 8.6 Dimensionless couple-stress distribution for $\alpha = 1.$; Eq. (8.77)

8.3.2 Dissipation

We recall that according to Eq. (6.23) the stress power in a Cosserat continuum can be split as follows,

$$P = \sigma_{(ij)} D_{ij} + 2t_i^* (\dot{\omega}_i - w_i) + \mu_{ik} K_{ik} \tag{8.78}$$

In the considered example we have,

$$P = 2\sigma_{(12)} D_{12} + 2t_3^* (\dot{\omega}_3 - w_3) + \mu_{23} K_{23} \tag{8.79}$$

where

$$2t_3^* = \sigma_{12} - \sigma_{21} \quad ; \quad \dot{\omega}_3 = \dot{\omega} = -\frac{1}{2} v' \tag{8.80}$$

Thus,

$$P = \mu \left(v'^2 + \alpha (v' + 2w)^2 + \beta (\ell w')^2 \right) \tag{8.81}$$

From Eqs. (8.73), (8.66) and (8.77) we have,

$$
\begin{aligned}
v' &= \frac{\tau}{\mu}\left(1 + \frac{\alpha}{1+\alpha} f^*\right) \\
v' + 2w &= \frac{1+2\alpha}{1+\alpha}\frac{\tau}{\mu} f^* \\
\ell w' &= \frac{1}{2}\beta\frac{\tau}{\mu}\ell f^{*\prime}
\end{aligned}
\tag{8.82}
$$

From these expressions and the exponential decay of the distribution function f^* away from the boundaries, Eq. (8.70), we obtain the following asymptotic expresion for the stress power inside the shear layer (Fig. 8.7),

$$
P \approx P^{(out)}\left(1 + 2\frac{\alpha}{1+\alpha}\left(e^{-y^*} + e^{-(H^*-y^*)}\right)\right) \quad ; \quad P^{(out)} = \frac{\tau^2}{\mu} \tag{8.83}
$$

We assume that all stress power is dissipated in heat. From the above discussion we conclude that the 1st term on the r.h.s. of Eq. (8.83) gives practically constant dissipation across the shear layer, whereas the contribution of the other terms is confined inside the two boundary layers. To see the effect of that, we derive first the heat equation starting from first principles.

We assume that the free energy of the fluid is only a function of the absolute temperature T, thus

$$
\frac{De}{Dt} = jc\frac{DT}{Dt} \tag{8.84}
$$

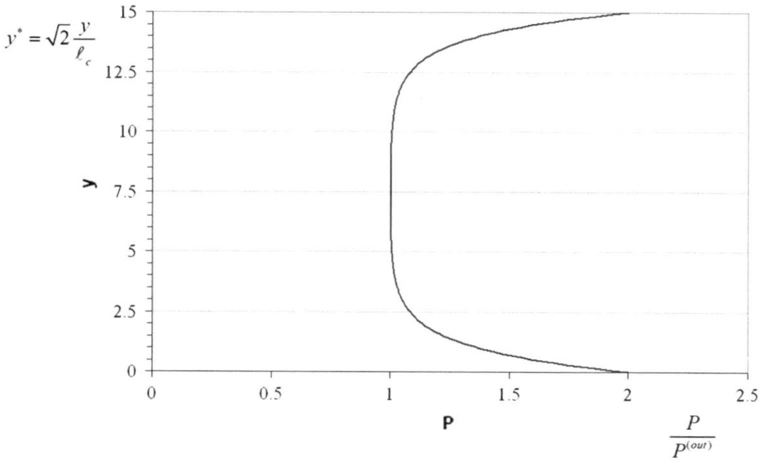

Fig. 8.7 Stress power distribution that is produced and is dissipated inside the Cosserat-fluid shear layer, for $\alpha = 1.$; Eq. (8.83)

where c is the specific heat of the medium and $j = 4.184 \frac{Joule}{cal}$ is the mechanical equivalent of heat, that appears in any equation that is related to the 1st Law and expresses the conversion of mechanical work into heat.

In addition, we adopt Fourier's law for isotropic heat conduction,

$$q_i = -jk_F \partial_i T \tag{8.85}$$

where k_F is the Fourier coefficient for heat conduction of the medium.

From the constitutive Eqs. (8.84) and (8.85) and the energy balance Eq. (6.24) we get,

$$\rho jc \frac{DT}{Dt} = P - j\partial_i (k_F \partial_i T) \tag{8.86}$$

For constant Fourier coefficient, the heat conduction equation for steady state conditions is,

$$\frac{d^2 T}{dy^2} + \frac{1}{jk_F} P = 0 \tag{8.87}$$

We scale again the $y-$ coordinate with the layer thickness

$$\bar{y} = \frac{y}{H} \tag{8.88}$$

and the temperature with the temperature at the boundaries, say T_0,

$$T^* = \frac{T}{T_0} \quad \Rightarrow \quad T^*(0) = T^*(1) = 1 \tag{8.89}$$

In terms of these dimensionless dependent variables the governing heat equation for steady heat production, Eq. (8.87), becomes

$$\frac{d^2 T^*}{d\bar{y}^2} + Gr\left(1 + 2\frac{\alpha}{1+\alpha}\left(e^{-y^*} + e^{-(H^* - y^*)}\right)\right) = 0 \tag{8.90}$$

were the dimensionless group,

$$Gr = \frac{\tau^2 H^2}{\mu jk_F T_0} \tag{8.91}$$

is called the *Gruntfest number*. The analytic solution of Eq. (8.90) is,

$$T^* = \bar{T}^* + \tilde{T}^* \tag{8.92}$$

where $\bar{T}^*$ is the classical parabolic solution holding also for classical (Boltzmann) Newtonian fluids,

$$\bar{T}^* = 1 + \frac{1}{2} Gr\bar{y}(1 - \bar{y}) \tag{8.93}$$

and $\tilde{T}^*$ is the superimposed boundary layer solution that is due to the Cosserat effects,

$$\tilde{T}^* = \frac{1}{2} Gr^* \left(\left(1 - e^{-y^*}\right) + \left(e^{-H^*} - e^{-(H^* - y^*)}\right) \right) \tag{8.94}$$

The dimensionless group Gr^* is a Gruntfest number that is refered to material length scale,

$$Gr^* = Gr \frac{2\alpha}{1+\alpha} \left(\frac{\ell}{H}\right)^2 = \frac{2\alpha}{1+\alpha} \frac{\tau^2 \ell^2}{\mu j k_F T_0} \tag{8.95}$$

This number governs also the extra heat that is produced at the boundary layers and will result into a net increase of the core temperature by,

$$\Delta T \approx \frac{1}{2} Gr^* T_0 = \frac{2\alpha}{1+\alpha} \frac{\tau^2 \ell^2}{\mu j k_F} \tag{8.96}$$

Note that if we set, $\ell \approx D_g$, then ΔT is in principle a measurable quantity. Note that according to Eq. (8.96) ΔT scales with the square of the particle size and the applied shear stress.

8.4 An Energy Consistent Granular Flow Model

8.4.1 Rate-Dependent Viscosity Functions

We consider the constitutive equations for an incompressible Cosserat fluid, discussed in Sect. 8.1, Eqs. (8.18). Based on this model we compute the stress power that is assumed to be totally dissipated in heat,

$$D_{loc} = P = \mu \Gamma^2 + \mu_c \Gamma_c^2 + c_T \mathrm{K}_T^2 + c_B \mathrm{K}_B^2 + c'_B \mathrm{K}_{B'}^2 \tag{8.97}$$

where we define the following invariants,

$$
\begin{aligned}
\Gamma^2 &= \Gamma_{ki}\Gamma_{ki} + \Gamma_{ki}\Gamma_{ik} = \Gamma_{ki}2\Gamma_{(ki)} = 2\Gamma_{(ki)}\Gamma_{(ki)} \geq 0 \\
\Gamma_c^2 &= \Gamma_{ki}\Gamma_{ki} - \Gamma_{ki}\Gamma_{ik} = \Gamma_{ki}2\Gamma_{[ki]} = 2\Gamma_{[ki]}\Gamma_{[ki]} \geq 0 \\
\mathrm{K}_T^2 &= \mathrm{K}_{mm}^2 \geq 0 \\
\mathrm{K}_B^2 &= \mathrm{K}_{ki}\mathrm{K}_{ki} \geq 0 \\
\mathrm{K}_{B'}^2 &= \mathrm{K}_{ki}\mathrm{K}_{ik} \geq 0
\end{aligned}
\tag{8.98}
$$

The last inequality follows directly from the polar decomposition of the distortion tensor: Assuming that K_{ij} is regular, $\left|\mathrm{K}_{ij}\right| \neq 0$, then its left polar decomposition reads

$$
[\mathrm{K}] = [S][Q] \quad ; \quad [S] = [S]^T \quad ; \quad |Q| = \pm 1 \tag{8.99}
$$

Thus

$$
[\mathrm{K}][\mathrm{K}]^T = [S][Q][Q]^T[S]^T = [S]^2 \quad \Rightarrow \quad tr[\mathrm{K}][\mathrm{K}]^T \geq 0 \tag{8.100}
$$

These remarks prompt to make the following constitutive assumptions concerning the viscosity functions,

$$
\begin{aligned}
&\mu = f_1(\Gamma) \quad ; \quad \mu_c = f_2(\Gamma_c) \\
&c_T = f_3(\mathrm{K}_T) \quad ; \quad c_B = f_4(\mathrm{K}_B) \quad ; \quad c_B' = f_5(\mathrm{K}_{B'})
\end{aligned}
\tag{8.101}
$$

such that,

$$
\mu = \frac{\partial D_{loc}}{\partial \Gamma} \quad , \quad etc.
$$

For the considered case of shear flow, we get from Eqs. (8.38) and (8.39) that

$$
\Gamma_s = |v'| \quad ; \quad \Gamma_c = |v' + 2w| \tag{8.102}
$$

and

$$
\mathrm{K}_T = \mathrm{K}_{B'} = 0 \quad ; \quad \mathrm{K}_B = |w'| \tag{8.103}
$$

Thus, an energy consistent model for rate-dependent viscosities in shear flows of an incompressible Cosserat fluid is the following,

$$
\mu = f_1(|v'|) \quad ; \quad \mu_c = f_2(|v' + 2w|) \tag{8.104}
$$

and

$$
c_B = f_4(|w'|) \tag{8.105}
$$

In order to make this model more realistic and closer to the theme of these Notes, we adopt here some findings that apply to *granular flows*.

The shearing resistance of a granular medium is pressure sensitive. From the point of view of a classical continuum the shear stress obeys formally a friction law,

$$\sigma_{(12)} = p \tan \varphi_s \tag{8.106}$$

where φ_s is the so-called *Coulomb friction angle* of the granular medium. In particular in shallow, rapid granular flows it is observed and/or computed using DEM that the friction is rate dependent [4]. The granular medium at rest possesses a *static friction coefficient*, $\tan \varphi_{st}$, while, if sheared at relative high shearing-rates, it gets fluidized and the internal friction increases from its static value to an asymptotic value, that we call the *dynamic friction coefficient*, $\tan \varphi_{dn} > \tan \varphi_{st}$ [7]. A simple law that interpolates between these two extreme values is an exponential law of the form,

$$\tan \varphi_s = \tan \varphi_{dn} - (\tan \varphi_{dn} - \tan \varphi_{st})\mathrm{e}^{-I_s} \tag{8.107}$$

where I_s is the *"inertial number"* of the flow, defined as

$$I_s = T_c|v'| > 0 \tag{8.108}$$

T_c is a characteristic time factor, that is reflecting inertia effects at grain scale,

$$T_c = \sqrt{\frac{D_g}{g}} \tag{8.109}$$

In the above definition D_g is the grain size and g is the acceleration of gravity.

Note that for small values of the inertial number, Eq. (8.107) is approximated by a linear law, as was suggested by Da Cruz et al. [8],

$$\tan \varphi_s = \tan \varphi_{st} + (\tan \varphi_{dn} - \tan \varphi_{st})I_s + O(I_s^2) \tag{8.110}$$

Equation (8.107) is a *visco-plastic* constitutive law, since rate effects become apparent above the given threshold placed by the static friction. In order to keep the model consistent with the assumptions that hold for a fluid, we adapt here this law so as to describe a *granular fluid* by setting $\tan \varphi_{st} = 0$, thus yielding the following constitutive equation for the *mobilized* friction coefficient,

$$\tan \varphi_s = \tan \varphi_{dn}\left(1 - \mathrm{e}^{-I_s}\right) \tag{8.111}$$

This assumption together with Eq. (8.106) leads to the following constitutive law for a pressure- and rate-dependent viscosity,

$$\mu = pT_c \tan \varphi_{dn} \mu^*(I_s) \tag{8.112}$$

where

$$\mu^*(I) = \frac{1 - e^{-I}}{I} \tag{8.113}$$

It is obvious that the constitutive Eq. (8.112) is consistent with the energy requirement, placed above by Eq. (8.104), since

$$I_s = T_c \Gamma_s \tag{8.114}$$

Moreover from this exercise we gain an insight about the probable form for the other two viscosity functions that are needed to close the problem. For example we may introduce two more inertial numbers,

$$I_c = T_c \Gamma_c = T_c |v' + 2w| \tag{8.115}$$

$$I_B = D_g T_c \mathrm{K}_B = D_g T_c |w'| > 0 \tag{8.116}$$

These inertial numbers together with the one originally defined for granular flows, Eq. (8.114), are proportional to the above discussed rate-of-deformation and distortion invariants, Eqs. (8.102) and (8.103). With this choice we may test as a minimal set the following viscosity functions,

$$\mu_c = pT_c \tan \varphi_{dn} \mu^*(I_a) \tag{8.117}$$

$$c_B = pT_c D_g^2 \tan \varphi_{dn} \mu^*(I_B) \tag{8.118}$$

In the next section we will apply this model to a standard steady shallow granular flow problem.

8.4.2 *Steady, Shallow Flow-Slide of a Granular Fluid*

We consider the problem of a steady gravitational flow of an incompressible Cosserat fluid, down an incline of infinite extend at constant slope angle θ (Fig. 8.8). As in Sect. 8.3 and in the previous subsection, we assume here that the only two significant kinematic variables of this problem are the particle velocity in the x_1- direction of the slope and the particle spin in the x_3- direction, perpendicular to the plane of the incline. Again, both fields are assumed to be functions only of $x_2 = y$, and accordingly the set of constitutive Eqs. (8.40) to (8.45), and the

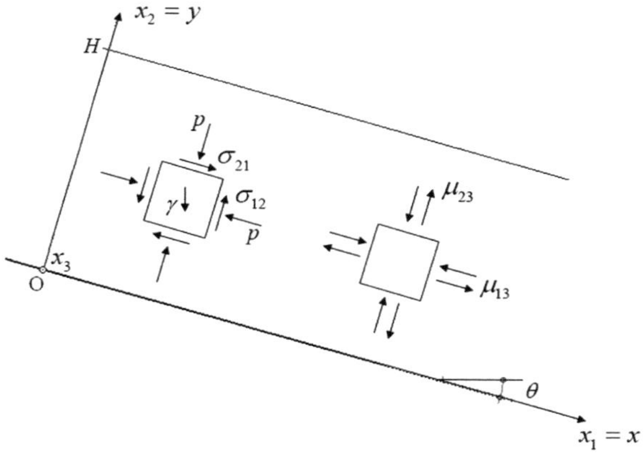

Fig. 8.8 Flow slide down an incline of infinite extend

moment stress equilibrium Eq. (8.48) also hold. Since we consider here a gravity-driven flow problem, we have to resort to the stress equilibrium Eqs. (4.42), which in the presence of gravity and for a granular fluid with density ρ yield,

$$\begin{aligned} \frac{\partial \sigma_{21}}{\partial y} + \rho g \sin\theta = 0 \quad &\Rightarrow \quad \sigma_{21} = \tau(0) - \rho g \sin\theta y \\ -\frac{\partial p}{\partial y} - \rho g \cos\theta = 0 \quad &\Rightarrow \quad p = p(0) - \rho g \cos\theta y \end{aligned} \tag{8.119}$$

The integration constants in Eqs. (8.119) follow directly from the condition that the upper boundary of the flow-slide at $y = H$ is stress-free; thus

$$\sigma_{21}(H) = 0 \quad \Rightarrow \quad \sigma_{21} = \rho g \sin\theta(H - y) \tag{8.120}$$

$$\sigma_{22}(H) = 0 \quad \Rightarrow \quad p = \rho g \cos\theta(H - y) \tag{8.121}$$

From the constitutive Eq. (8.42), the above introduced constitutive assumtions for the rate-dependent viscosities and Eq. (8.120) we get,

$$\mu^*(I_s)T_c v' + \mu^*(I_a)T_c(v' + 2w) = \lambda(\theta) \tag{8.122}$$

where

$$\lambda = \frac{\tan\theta}{\tan\varphi_{dn}} < 1 \tag{8.123}$$

We assume that

$$v' > 0 \quad , \quad v' + 2w > 0 \quad , \quad w' > 0 \tag{8.124}$$

These assumptions must be checked a posteriori (Figs. 8.2, 8.4 and 8.6). By utilizing these inequalities, we get from Eqs. (8.122), (8.113) and the equilibrium Eqs. (8.48), (8.43) and (8.45) the following non-linear ordinary differential equations for the kinematic variables of the problem at hand,

$$\left(1 - e^{-T_c v'}\right) + \left(1 - e^{-T_c(v' + 2w)}\right) = \lambda \tag{8.125}$$

$$D_g \frac{1}{p} \frac{d}{dy} \left(p\left(1 - e^{-D_g T_c w'}\right)\right) - 2\left(1 - e^{-T_c(v' + 2w)}\right) = 0 \tag{8.126}$$

Next we try some analytical treatment of this set of equations.

8.4.2.1 Outer Approximation

Equation (8.126) describes again a boundary layer problem. The outer approximation of the solution for this problem follows from this equation by neglecting the term that is multiplied by D_g, thus yielding

$$1 - \exp\left(-T_c\left(v^{(out)\prime} + 2w^{(out)}\right)\right) = 0 \quad \Rightarrow \quad w^{(out)} = -\frac{1}{2} v'^{(out)} \tag{8.127}$$

cf. Equation (8.60). This result and Eq. (8.125) yield,

$$1 - \exp\left(-T_c v^{(out)\prime}\right) = \lambda \quad \Rightarrow \quad v^{(out)\prime} = \frac{1}{T_c} \ln\left(\frac{1}{1 - \lambda}\right) \tag{8.128}$$

Thus, as expected, the outer solution is again linear across the shear-layer. If we utilize the definition of the time factor T_c, Eq. (8.109), we get the following expression for the velocity gradient across the flow slide,

$$\frac{v_H}{H} \approx \sqrt{\frac{g}{D_g}} F_0(\lambda) \tag{8.129}$$

where v_H is the flow-velocity v_H at the top of the flow-slide, and F_0 is a function of the parameter λ,

$$F_0 = -\ln(1 - \lambda) \tag{8.130}$$

The parameter $0 < \lambda < 1$, depends according to Eq. (8.123) on the dynamic friction and on the slope angle. We note however that for $\lambda = 1$ $(\theta = \varphi_{dn})$ the present model breaks down.

As is done in the pertinent literature, the flow velocity is expressed in terms of the Froude number that refers to the grain size [7],

$$\mathrm{Fr_D} = \frac{v_H}{\sqrt{gD_g}} \tag{8.131}$$

In this case Eq. (8.129) yields a linear relation between Fr_D and the dimensionless flow-height, normalized by the grain size,

$$\mathrm{Fr_D} \approx F_0(\lambda)\frac{H}{D_g} \tag{8.132}$$

8.4.2.2 Shallow Flow-Slides

In order to investigate the effect of the forming boundary layer at the base of the flow slide, we return to the moment equilibrium Eq. (8.126). With,

$$\frac{1}{p}\frac{dp}{dy} = -\frac{1}{H-y} \tag{8.133}$$

in a first step of approximation this equation yields,

$$D_g^2\left(T_c w'' - \frac{1}{H-y}T_c w'\right) + 2\left(1 - e^{-T_c v'} - \lambda\right) = 0 \tag{8.134}$$

Equation (8.125) can be solved for w, yielding

$$T_c w = -\frac{1}{2}\ln((2-\lambda)e^{-q} - 1) \approx -\frac{1}{2}\left(\ln(1-\lambda) + \frac{2-\lambda}{1-\lambda}q\right) \tag{8.135}$$

where,

$$q = T_c v' \tag{8.136}$$

With this notation from Eq. (8.134) we get,

$$-\ell_c^2\left(q'' - \frac{1}{H-y}q'\right) + 2(1 - e^{-q} - \lambda) = 0 \tag{8.137}$$

where,

$$\ell_c = \sqrt{\frac{1}{2}\frac{2-\lambda}{1-\lambda}}D_g \tag{8.138}$$

We rescale the $y-$ coordinate according to Eq. (8.61) and we set

$$s = H^* - y^* \tag{8.139}$$

With this transformation the governing Eq. (8.137) becomes,

$$\frac{d^2q}{ds^2} + \frac{1}{s}\frac{dq}{ds} + e^{-q} - (1 - \lambda) = 0 \tag{8.140}$$

We set,

$$q = q^{(out)} + \tilde{q} \tag{8.141}$$

such that,

$$e^{-q^{(out)}} = 1 - \lambda \quad \Leftrightarrow \quad q^{(out)} = -\ln(1 - \lambda) \tag{8.142}$$

cf. Equation (8.128). Thus

$$\frac{d^2\tilde{q}}{ds^2} + \frac{1}{s}\frac{d\tilde{q}}{ds} - (1 - \lambda)\left(1 - e^{-\tilde{q}}\right) = 0 \tag{8.143}$$

Its linearized version,

$$\frac{d^2\tilde{q}}{ds^2} + \frac{1}{s}\frac{d\tilde{q}}{ds} - (1 - \lambda)\tilde{q} = 0 \tag{8.144}$$

admits an analytic solution in terms of modified Bessel functions,

$$q = C_1 I_0(\sqrt{1 - \lambda} s) + C_2 K_0(\sqrt{1 - \lambda} s) \tag{8.145}$$

At the top of the flow-slide $(y = H)$ we assume that the couple stress μ_{23} is zero, which according to Eqs. (8.43), (8.135) and (8.139) yields

$$y = H \Rightarrow \quad s = 0 : w' = 0 \quad \Rightarrow \quad \left.\frac{dq}{ds}\right|_{s=0} = 0 \tag{8.146}$$

With,

$$\frac{dq}{ds} = -C_1 K_1\left(\sqrt{1 - \lambda} s\right) + C_2 I_1\left(\sqrt{1 - \lambda} s\right) \tag{8.147}$$

the condition at the free boundary leads to $C_1 = 0$, thus,

$$\tilde{q} = C_2 I_0\left(\sqrt{1 - \lambda} s\right) \tag{8.148}$$

At the base of the flow-slide ($y = 0$) we apply the no-spin boundary condition, which according to Eqs. (8.135) and (8.136) reads,

$$\begin{aligned} y = 0: \quad w = 0 \Rightarrow \quad q = q^{(out)} + \tilde{q} = -\frac{1-\lambda}{2-\lambda}\ln(1-\lambda) \Rightarrow \\ \tilde{q} = \frac{1}{2-\lambda}\ln(1-\lambda) \end{aligned} \tag{8.149}$$

This gives,

$$\tilde{q} = \frac{1}{2-\lambda}\ln(1-\lambda)\frac{I_0\left(\sqrt{1-\lambda}(H^* - y^*)\right)}{I_0(\sqrt{1-\lambda}H^*)} \tag{8.150}$$

and with that,

$$\begin{aligned} q &= -\ln(1-\lambda)\left(1 - \frac{1}{2-\lambda}\frac{I_0\left(\sqrt{1-\lambda}(H^* - y^*)\right)}{I_0\left(\sqrt{1-\lambda}H^*\right)}\right) \\ &\approx -\ln(1-\lambda)\left(1 - \frac{1}{2-\lambda}e^{-\sqrt{1-\lambda}y^*}\right) \end{aligned} \tag{8.151}$$

From Eq. (8.151) we get

$$\frac{dv^*}{dy^*} = q(y^*) \quad ; \quad v^* = \sqrt{2}\frac{T_c}{\ell_c}v \tag{8.152}$$

The solution for the velocity is obtained by integrating Eq. (8.152) and by utilizing the non-slip condition at the base of the flow-slide,

$$y = 0: \quad v = 0 \tag{8.153}$$

The predicted solution corresponds to concave/linear velocity profile across the flow-slide, as is shown in Fig. 8.9,

$$v^* = -\ln(1-\lambda)\left(1 - \frac{1}{2-\lambda}\frac{1 - e^{-\sqrt{1-\lambda}y^*}}{\sqrt{1-\lambda}y^*}\right)y^* \tag{8.154}$$

From Eq. (8.154) we get further the following relation between Froude number and flow-height,

$$\mathrm{Fr_D} \approx F_1\left(\lambda, \frac{H}{D_g}\right)\frac{H}{D_g} \tag{8.155}$$

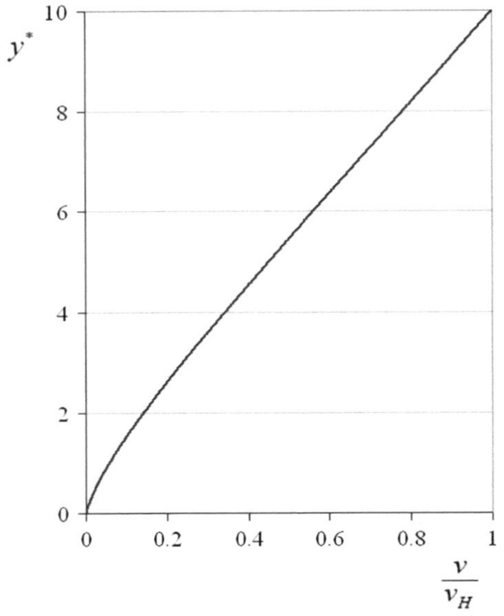

Fig. 8.9 Predicted locally concave/linear velocity profile across the flow-slide thickness, ($\lambda = 0.5$); Eq. (8.154)

where

$$F_1 = -\ln(1-\lambda)\left(1 - \frac{1}{2-\lambda}\frac{1-e^{-\sqrt{2-\lambda}\frac{H}{D_g}}}{\sqrt{2-\lambda}\frac{H}{D_g}}\right) \tag{8.156}$$

We note that for large values of H/D_g, Eq. (8.155) tends to a line that is parallel to the one we obtained by utilizing the outer approximation. Deviations from linearity hold for relatively small values for H/D_g, where the boundary layers influence the solution. In Fig. 8.10 we compare the outer solution, Eq. (8.132) and the Cosserat-approximate solution, Eq. (8.155). From this figure one can see clearly that consideration of the boundary layer that is forming at the base of the flow-slide provides more conservative estimates for the top flow-velocity. It is however important to notice that for high values of the Froude number, the steady solution is linearly unstable and that stability must be investigated before adopting such a result.

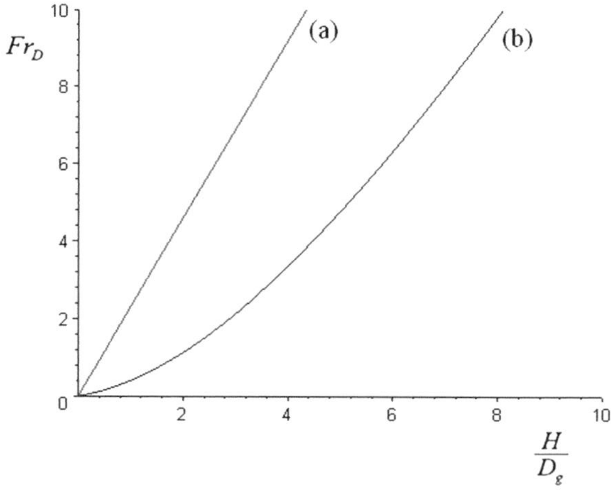

Fig. 8.10 Normalized flow-velocity as function of the normalized flow-height. Comparison **a** classical (outer) solution. **b** Cosserat solution ($\lambda = 0.9$)

8.4.2.3 Exercise: Flow-Slide with Basal Slippage

If we evaluate the basal stresses and couple-stress from the previously analyzed problem, they take the following form,

$$\begin{aligned} \sigma_{22}(0) &= -\sigma_0 = -p(0) = -\rho g H \cos\theta \\ \sigma_{21}(0) &= -\tau_0 = \tau(0) = \rho g H \sin\theta \\ \mu_{23}(0) &= -m_0 = -\tau_0 \ell' \end{aligned} \tag{8.157}$$

As can be seen from Fig. 8.11, the appearance of the couple stress at the base of the flow-slide may be interpreted as a surface roughness effect [9]. This can be envisioned by letting the shear stress τ_0 to be applied at a distance ℓ' away from the theoretical base and its effect being transported to the base through a series of rigid ortho-fibers that are continuously attached to the basal surface.

Following these remarks, the previously discussed problem can be solved for an alternative set of boundary conditions at the base of the slide, that will allow in turn for slippage to occur. First we prescribe the basal couple stress as function of the basal shear stress, Eq. (8.157). This is a Neumann-type boundary condition for the particle spin. Thus, it allows for the particles to rotate at the basal plane. In this case it is not physically sound to assume the non-slip condition, since the rigid-body rotation of the particle that finds itself at the basal plane will impose a velocity on that particle (Fig. 8.12),

$$v(0) = -\frac{D_g}{2} w(0) \tag{8.158}$$

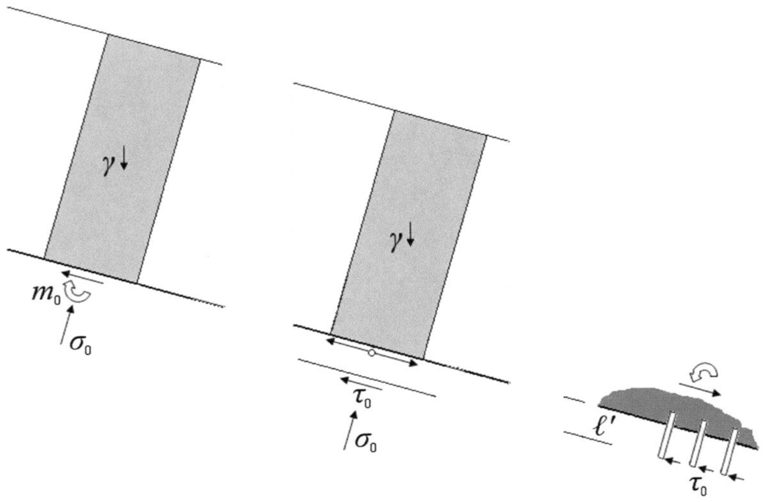

Fig. 8.11 Mechanical interpretation of the basal stress-and couple-stress boundary conditions as a surface roughness effect

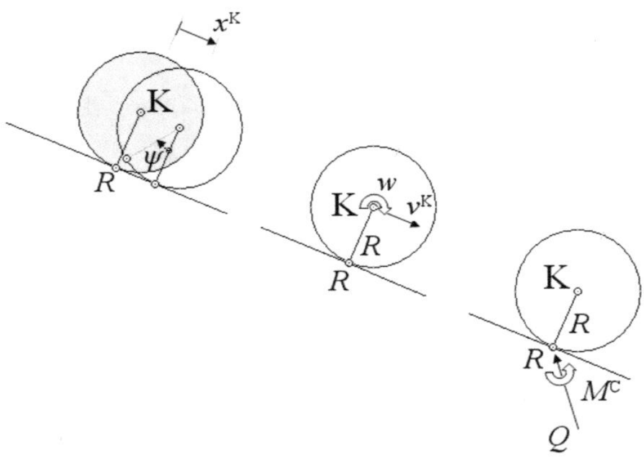

Fig. 8.12 Sphere rolling down a planar incline. The velocity of the sphere is directly linked to its rotation. Basal friction may include rolling resistance due to existence of asperities and other surface roughness agents

Let,

$$\lambda' = \lambda \frac{\ell'}{D_g} < \lambda \tag{8.159}$$

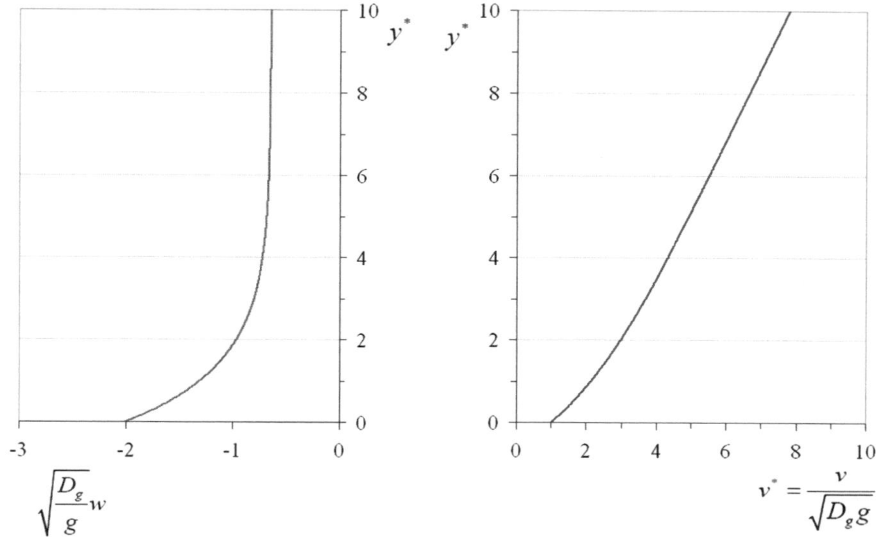

Fig. 8.13 Particle spin profile with basal rolling and convex particle velocity profile with basal slippage ($\lambda = \lambda' = 0.485$); Eqs. (8.161) and (8.162)

As an exercise one is asked here to determine the solution of this boundary value problem in terms of particle velocity and spin. Using the notation of the previous section, one should check as intermediate results that the following equations hold,

$$q \approx -\ln(1-\lambda) - \frac{\ln(1-\lambda')}{\sqrt{1-\lambda}}\exp\left(-\sqrt{1-\lambda}y^*\right) \tag{8.160}$$

$$T_c w \approx \frac{1}{2}\frac{\ln(1-\lambda)}{1-\lambda} + \frac{1}{2}\frac{2-\lambda}{1-\lambda}\frac{\ln(1-\lambda')}{\sqrt{1-\lambda}}\exp\left(-\sqrt{1-\lambda}y^*\right) \tag{8.161}$$

$$\frac{dv^*}{dy^*} = \frac{1}{2}\sqrt{\frac{2-\lambda}{1-\lambda}}q(y^*) \tag{8.162}$$

Compare your results with the ones depicted in Fig. 8.13.

References

1. Condiff, D. W., & Dahler, J. S. (1964). Fluid mechanical aspects of antisymmetric stress. *Physics of Fluids, 7*(6), 842–854.
2. Eringen, A. C. (1964). Simple microfluids. *International Journal of Engineering Science, 2*, 205–217.

3. Rao, L. (1970). Stability of micropolar fluid motions. *International Journal of Engineering Science, 8,* 753–762.
4. Mitarai, N., Hayakawa, H., & Nakanishi, H. (2002). Collisional granular flow as a micropolar fluid. *Physical Review Letters, 88,* 174301.
5. Hinch, E. J. (1991). *Perturbation methods*. Cambridge University Press.
6. Vardoulakis, I., & Sulem, J. (1995). *Bifurcation analysis in geomechanics*. Blackie Academic & Professional.
7. Pouliquen, O. (1999). Scaling laws in granular flows down rough inclined planes. *Physics of Fluids, 11,* 542–548.
8. da Cruz, F., et al. (2005). Rheophysics of dense granular materials: discrete simulation of plane shear flows. *Physical Review E, 72,* 021309.
9. Vardoulakis, I., Shah, K. R., & Papanastasiou, P. (1992). Modeling of tool-rock interfaces using gradient dependent flow-theory of plasticity. *International Journal of Rock Mechanics and Mining Sciences & Geomechanics Abstracts, 29,* 573–582.

Chapter 9
Mechanics of Discrete Granular Media

Abstract This chapter links the Cosserat continuum with discrete granular media. Through a discrete modelling approach, it presents a homogenisation method based on intergranular energetics and fabric averaging.

We consider here the basic statics and kinematics of discrete media, consisting of rigid grains in semi-permanent contact. The usually irregular in shape grains will be pictured as circles and the mutual grain (multiple) contacts as single point contacts. This picture retains the main topological properties of the thought configuration of a granular medium. Knowledge concerning shape and size of grains, the nature of their contacts and the description of their packing in 3D space will always be incomplete. Having this in mind, we try here to not overdo with graph-theoretical and statistical physics considerations, and notations as well as the related jungle of assumptions. In principle we will try to keep our set of assumptions to a bare minimum. In this section we use Cartesian notation throughout.

9.1 Compatibility in the Discrete Setting

Let us consider two particles (p_α) and (p_β) in contact at point P_c (Fig. 9.1). We denote with $\boldsymbol{w}^{(\alpha)}$ the spin of the particle (p_α) and with $\boldsymbol{v}^{\mathrm{K}_\alpha}$ the velocity of the centroid K_α of particle (p_α). All points of (p_α) including the contact point P_c share its rigid body motion, thus their motion is given by the corresponding kinematic particle motor,

$$\bar{k}^{(\alpha)} = \begin{pmatrix} \boldsymbol{w}^{(\alpha)} \\ \boldsymbol{v}^{\mathrm{P}_c} \end{pmatrix} \tag{9.1}$$

such that

$$\boldsymbol{w}^{(\alpha)\,\mathrm{K}_\alpha} = \boldsymbol{w}^{(\alpha)\,\mathrm{P}_c} \tag{9.2}$$

I. Vardoulakis, *Cosserat Continuum Mechanics*, Lecture Notes in Applied and Computational Mechanics 87, https://doi.org/10.1007/978-3-319-95156-0_9

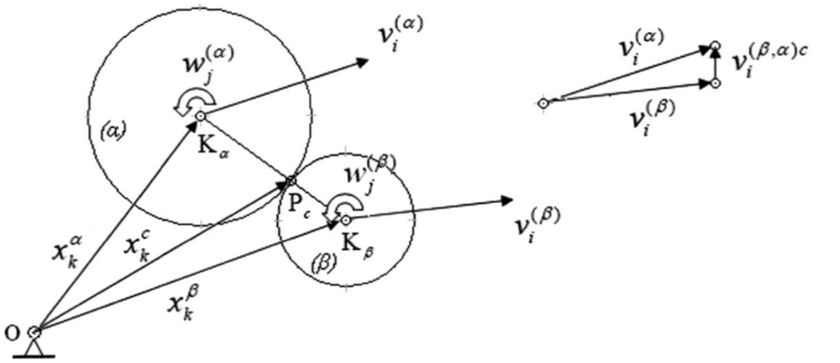

Fig. 9.1 Relative velocity of two grains in contact

$$\boldsymbol{v}^{P_c} = \boldsymbol{v}^{K_\alpha} + \boldsymbol{w}^{(\alpha)} \times \overrightarrow{K_\alpha P_c} \tag{9.3}$$

We observe that this property of the discrete particles is in harmony with local Cosserat continuum kinematics, as this is expressed above by Eqs. (3.8) and (3.9).

In Cartesian components Eq. (9.3) reads

$$v_i^{(\alpha)c} = v_i^{(\alpha)} + \varepsilon_{ijk} w_j^{(\alpha)} \left(x_k^c - x_k^\alpha\right) \tag{9.4}$$

Similarly for particle p_β we have,

$$v_i^{(\beta)c} = v_i^{(\beta)} + \varepsilon_{ijk} w_j^{(\beta)} \left(x_k^c - x_k^\beta\right) \tag{9.5}$$

These expressions allow us to compute the relative motion between two neighboring particles in contact; i.e. their relative spin

$$w_i^{(\beta,\alpha)} = w_i^{(\beta)} - w_i^{(\alpha)} \tag{9.6}$$

and their relative velocity at the contact point,

$$\begin{aligned} v_i^{(\beta,\alpha)c} &= v_i^{(\beta)c} - v_i^{(\alpha)c} \Rightarrow \\ v_i^{(\beta,\alpha)c} &= v_i^{(\beta)} - v_i^{(\alpha)} + \varepsilon_{ijk}\left(w_j^{(\beta)}\left(x_k^c - x_k^\beta\right) - w_j^{(\alpha)}\left(x_k^c - x_k^\alpha\right)\right) \end{aligned} \tag{9.7}$$

Let us consider an open line of N-grains in sequential contact, usually termed also a *granular column* (Fig. 9.2). If we apply Eqs. (9.6) and (9.7) consecutively, then we get the following expressions for the difference in rotation and displacement between two grains in “remote” contact,

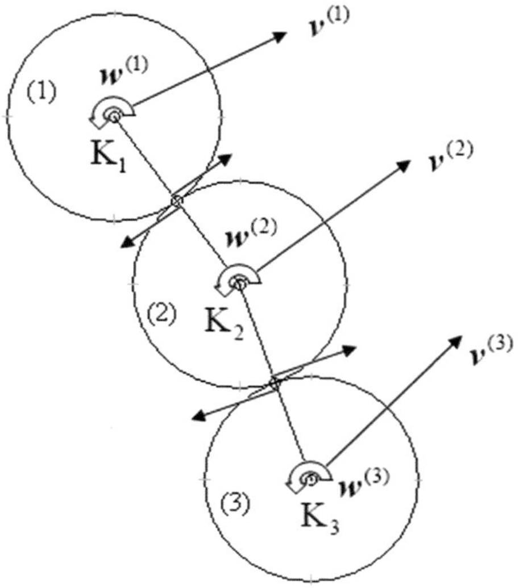

Fig. 9.2 Open line of homothetically rotating grains

$$\Delta w_i^{(N,1)} = w_i^{(N)} - w_i^{(1)} = \sum_{\alpha=1}^{N-1} w_i^{(\alpha,\alpha+1)} \tag{9.8}$$

and

$$\Delta v_i^{(N,1)} = u_i^{(N)} - u_i^{(1)} = \sum_{\alpha=1}^{N-1} \left(v_i^{(\alpha,\alpha+1)c} - \varepsilon_{ikj} \left(x_k^{c(\alpha+1,\alpha)} - x_k^{c(\alpha,\alpha-1)} \right) w_j^{(\alpha)} \right) \tag{9.9}$$

where $x_k^{(1,0)c} \equiv x_k^1$ and $x_k^{(N,N+1)c} \equiv x_k^N$.

Following a remark by Satake [1], Eqs. (9.8) and (9.9) are seen as the discrete manifestation of the line integrals, Eqs. (3.99) and (3.105), holding for a Cosserat continuum:

$$\begin{aligned} \Delta \psi^{i(P_2,P_1)} &= \psi^i(P_2) - \psi^i(P_1) = \int_{P_1}^{P_2} \kappa_k^{\cdot i} d\Theta^k \leftrightarrow \Delta w_i^{(N,1)} = w_i^{(N)} - w_i^{(1)} \\ &= \sum_{\alpha=1}^{N-1} w_i^{(\alpha,\alpha+1)} \end{aligned} \tag{9.10}$$

and

$$\begin{aligned} &\Delta u_i^{(P_2,P_1)} = u_i(P_2) - u_i(P_1) = \int_{P_1}^{P_2} \left(\gamma_{ik} - e_{ikl}\psi^l\right) d\Theta^k \\ &\leftrightarrow \\ &\Delta \delta v_i^{(N,1)} = \delta u_i^{(N)} - \delta u_i^{(1)} = \sum_{\alpha=1}^{N-1} \left(\delta v_i^{(\alpha,\alpha+1)c} - \varepsilon_{ikj} \left(x_k^{c(\alpha+1,\alpha)} - x_k^{c(\alpha,\alpha-1)} \right) \delta w_j^{(\alpha)} \right) \end{aligned} \tag{9.11}$$

With the symbol $\leftrightarrow$ we depict here an observed analogy between the continuum- and the discrete mathematical description. Satake's analogy that is displayed above allows us to identify:

1. The Cosserat continuum rotation as that kinematical property of the continuum that is meant to reproduce the particle rotation.
2. The relative deformation of the Cosserat continuum as measure for relative or inter-particle displacement.
3. The distortions of the Cosserat continuum as the measure for the relative inter-particle rotation.

The discrete and the continuous realization of the relative displacement and relative rotation between two points are given by Eqs. (9.10) and (9.11). If these relative motions are path independent, then we are dealing with "compatible" deformations. In particular the relative motions in a compatible deformation should vanish, if evaluated in a closed loop. This is not always the case in granular media. To demonstrate this statement we consider the paradigm of the planar, 3-grain circuit of Fig. 9.3.

For simplicity we assume that the "grains" in Fig. 9.3 are indeed circular rods of equal radius R_g and that grains (1) and (2) are spinning homothetically, grain (3) is spinning antithetically, all with the same strength $\delta\omega$. We see immediately that this constellation provides two pure rolling contacts (rc) at c_1 and c_3, and a pure sliding contact (sc) at c_2. We note that the relative displacement between two neighboring grains is null across pure rolling contacts. For the virtual motion of this circuit that is shown in Fig. 9.3 we compute,

$$\begin{aligned} &\Delta\delta w_3^{(2,1)} = 0; \quad \Delta\delta\omega_3^{(1,3)} = 2\delta w; \quad \Delta\delta w_3^{(3,2)} = -2\delta w \\ \Rightarrow \quad &\sum_{cycl} \Delta\delta w_3^{(\alpha,\beta)} = 0 \end{aligned} \tag{9.12}$$

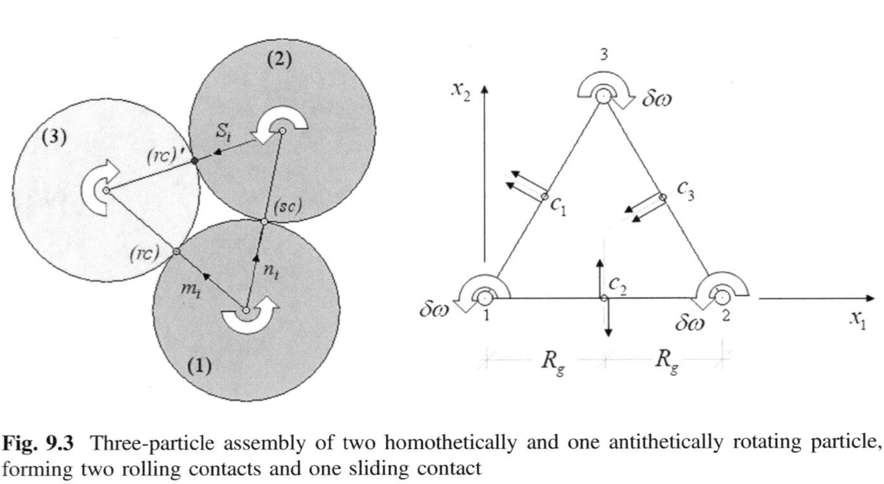

Fig. 9.3 Three-particle assembly of two homothetically and one antithetically rotating particle, forming two rolling contacts and one sliding contact

and

$$
\begin{aligned}
&\Delta\delta v_2^{(2,1)c} = R_g\delta w; \quad \Delta\delta u_2^{(1,3)c} = R_g\delta w; \quad \Delta\delta v_2^{(3,2)c} = -R_g\delta w \\
&\Rightarrow \quad \sum_{cycl} \Delta\delta v_2^{(\alpha,\beta)c} = R_g\delta w \neq 0
\end{aligned}
\tag{9.13}
$$

$$
\begin{aligned}
&\Delta\delta v_1^{(2,1)c} = 0; \quad \Delta\delta v_1^{(1,3)c} = -\tfrac{\sqrt{3}}{4}\delta w; \quad \Delta\delta v_1^{(3,2)c} = -\tfrac{\sqrt{3}}{4}R_g\delta w \\
&\Rightarrow \sum_{cycl} \Delta\delta v_2^{(\alpha,\beta)c} = -\tfrac{\sqrt{3}}{2}R_g\delta w \neq 0
\end{aligned}
\tag{9.14}
$$

This means that incorporation of antithetically rotating particles into our consideration, would mean to extend the Cosserat model to incompatible deformations, as was the case for example in Günther's [2] interpretation of Kröner's [3] theory of dislocations. Note that in the terminology of granular Physics, if the grains in a network can rotate without sliding on each other, then the network is called *dynamically unfrustrated*, since it can deform freely under shear and behaves like a dry fluid. A sufficient condition for "non-frustration" in 3D is that all closed circuits of grains in contact are even [4]. This means in turn that solid granular matter differs from granular fluid in the aspect that in solid granular mater sliding is the rule and rolling the exception. What we call in granular solid Mechanics *compatibility* is called in granular fluid Physics *frustration*. As we will discuss next, the transition from a fully *frustrated* to a partially *unfrustrated* system is an instability, that is related in granular Mechanics with *shear-banding*. This observation allows also to view shear banding as a granular solid-fluid *phase transition*. To this end we recall an early statement by Oda and Kazama [5], who remarked that: "… that a shear band grows through buckling of columns together with rolling at contacts; it can be said that the thickness of a shear band is determined by the number of particles involved in a single column". Indeed, from the micro-mechanical point of view an important structure that appears to dominate localized deformation in 2D DEM simulations is the formation and collapse (*buckling*) of grain columns, as this was demonstrated experimentally by Oda and was given a simple structural mechanics description by Satake [6]. These load-carrying columns belong to the so-called *competent grain fraction*, a concept first introduced by Dietrich [7] and later used in continuum shear-banding analyses by Vardoulakis [8]. In later years the existence of the *bimodal* character of the contact-forces network in granular media was filtered-out from numerical CD simulations, by Radjai and co-workers [9, 10]. The length of these buckling granular columns reflects more or less the current shear band thickness. Recently Tordesillas [11] picked on this matter and pointed that "One such unjamming mechanism is the buckling of force chains and associate growth of surrounding voids … This mechanism is characteristically non-affine". The term *non-affine* in connection to an open line of grains is, in our understanding, not outside Günther's original idea of incompatible deformations and Satake's integrability and dislocation concepts. This can be seen in Fig. 9 of Ref. [11], where we observe that the line of grains that caries *non-affine deformation* information includes antithetically rotating grains. To our understanding, the essential feature here is the incompatibility of grain rotation across a grain contact that is leading to the possibility of an internal instability in the form of the total plastic yielding of an internal (frictional) hinge. Thus, shear-banding as the manifestation of an internal instability in the sense of Oda should include the formation of plastic hinges between grains that belong to the strong force network. Note that in terms of continuum mechanics, the formation of plastic hinges aligned more or less in a *granular surface* [12] inside a shear band, means the appearance of finite jumps (i.e. strong discontinuities) in the particle spin.

9.2 Equilibrium in the Discrete Setting

9.2.1 Definitions

Following Bardet and Vardoulakis [13] we consider a Representative Elementary Volume (REV) that consists of N sub-particles ("grains"), some of which are subjected to external forces or couples, applied from the exterior of the considered (REV); Fig. 9.4. The particles inside the (REV) are grouped in the set $B = \{p_\alpha | \alpha = 1, \ldots, N\}$. The forces and couples acting on the particles of B are reduced at M points that form the set of "contact" points, $C = \{\mathrm{P}_c | c = 1, \ldots, M\}$. The subset $I \subset C$ contains the contact points between two particles of B, whereas the subset $E \subset C$ contains the points where external actions are applied,

$$\begin{aligned} &I = \{\mathrm{P}_1, \ldots, \mathrm{P}_{M_I}\}, \quad E = \left\{\mathrm{P}_{M_{I+1}}, \ldots, \mathrm{P}_M\right\} \\ &C = I \cup E, \quad \varnothing = I \cap E \end{aligned} \tag{9.15}$$

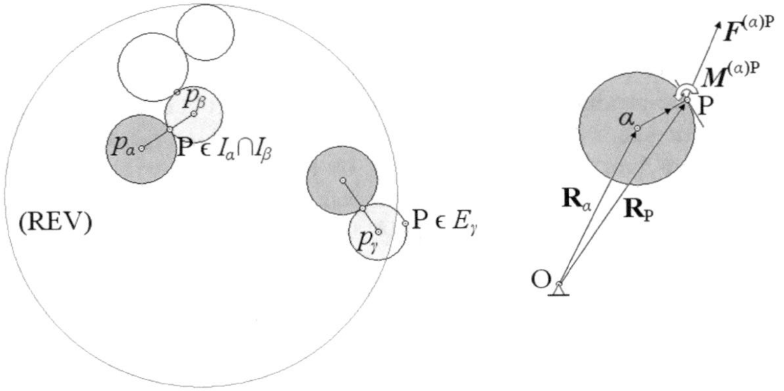

Fig. 9.4 (REV) containing set B of particles in contact among each other and other exterior particles

Sets I_α, E_α and C_α denote the contact points on particle p_α corresponding to internal actions, external actions and all actions. Sets C_α, I_α, E_α, I, E and C are related as follows:

$$C = \bigcup_{p_\alpha \in B} C_\alpha, \quad C_\alpha = I_\alpha \cup E_\alpha$$
$$I = \bigcup_{p_\alpha \in B} I_\alpha \tag{9.16}$$
$$E = \bigcup_{p_\alpha \in B} E_\alpha$$

The intersections of I_α and E_α are either empty or reduced to a single (contact) point,

$$\begin{aligned} E_\alpha \cap E_\beta &= \varnothing \quad (\alpha \neq \beta) \\ I_\alpha \cap I_\beta &= \{c_i \in I\}.or.\varnothing \quad (\alpha \neq \beta) \end{aligned} \tag{9.17}$$

9.2.2 Equilibrium of the Single Particle

The mechanical action on particle (p_α) from one of its neighbors is reduced to the contact action motor which consists of a line force vector and a couple, that are transported at the "contact" point P_c (Fig. 9.5),

$$\underline{F}^{(\alpha)c} = \begin{pmatrix} \boldsymbol{F}^{(\alpha)\mathrm{P}_c} \\ \boldsymbol{M}^{(\alpha)\mathrm{P}_c} \end{pmatrix} \tag{9.18}$$

This is because particles in granular assemblies are not necessarily convex and they interlock. As explained by Froiio et al. [12], the selection of the "contact" point P_c is rather arbitrary, a fact that is very well reflected in the transport properties of the contact action motor, Eq. (9.18).

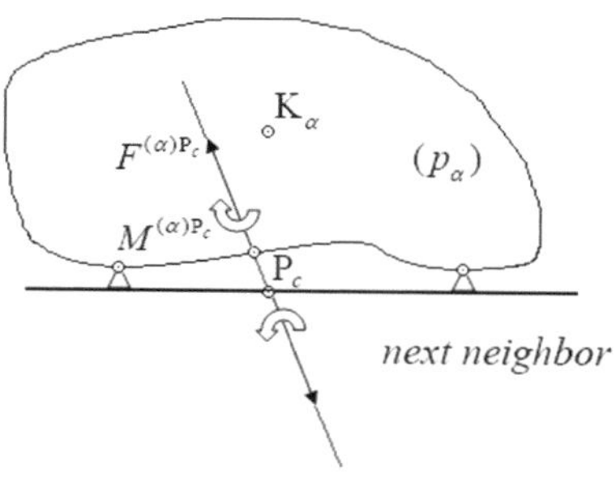

Fig. 9.5 The intergranular contact motor

For easy computation of the resulting forces and couples we may use again Cartesian notation for the related vectors

$$\boldsymbol{F}^{(\alpha)\mathrm{P}_c} = f_i^{(\alpha)c}\boldsymbol{e_i} \tag{9.19}$$

$$\boldsymbol{M}^{(\alpha)\mathrm{P}_c} = m_i^{(\alpha)c}\boldsymbol{e_i}, \quad m_i^{(\alpha)c} \equiv m_i^{(\alpha)}(\mathrm{P}_c) \tag{9.20}$$

The resultant of force acting on particle (p_α) is

$$\boldsymbol{F}^{(\alpha)} = \sum_{c \in \boldsymbol{C}_\alpha} \boldsymbol{F}^{(\alpha)\mathrm{P}_c} \tag{9.21}$$

We transport all forces and moments acting on (p_α) to its "centre" K_α[1], thus obtaining the total particle motor,

$$\underline{F}^{(\alpha)} = \begin{pmatrix} \boldsymbol{F}^{(\alpha)} \\ \boldsymbol{M}^{(\alpha)\mathrm{K}_\alpha} \end{pmatrix} \tag{9.22}$$

with

$$\boldsymbol{F}^{(\alpha)} = f_i^{(\alpha)}\boldsymbol{e_i}, \quad f_i^{(\alpha)} = \sum_{c \in C_\alpha} f_i^{(\alpha)c} \tag{9.23}$$

$$\boldsymbol{M}^{(\alpha)\mathrm{K}_\alpha} = m_i^{(\alpha)}\boldsymbol{e_i} \tag{9.24}$$

$$m_i^{(\alpha)} \equiv m_i^{(\alpha)}(\mathrm{K}_\alpha) = \sum_{c \in C_\alpha} \left(m_i^{(a)c} + \varepsilon_{ijk}\left(x_j^c - x_j^\alpha\right) f_k^{(\alpha)c} \right) \tag{9.25}$$

[2]and the position vectors

$$\mathbf{R}_{\mathrm{K}_\alpha} = x_i^\alpha \boldsymbol{e_i}, \quad \mathbf{R}_{\mathrm{P}_c} = x_i^c \boldsymbol{e_i} \tag{9.26}$$

Equilibrium for particle (p_α) is expressed by

$$\underline{F}^{(\alpha)} = \underline{0} \tag{9.27}$$

Let $\delta\bar{k}^{(\alpha)}$ be the kinematic motor for a virtual motion of the particle (p_α),

$$\delta\bar{k}^{(\alpha)} = \begin{pmatrix} \delta\boldsymbol{w}^{(\alpha)} \\ \delta\boldsymbol{v}^{\mathrm{K}_\alpha} \end{pmatrix} \tag{9.28}$$

---

[1]Point K_α could be the centroid of the grain.

[2]Note that in this equation summation of repeated lower indices is meant!

where

$$\delta \boldsymbol{w}^{(\alpha)} = \delta w_i^{(\alpha)} \boldsymbol{e_i}, \quad \delta \boldsymbol{v}^{\mathrm{K}_\alpha} = \delta v_i^{(\alpha)} \boldsymbol{e_i} \tag{9.29}$$

are the virtual spin vector of the particle and the virtual velocity vector of the centre K_α of particle p_α, respectively.

The virtual power of the total force and total couple acting on particle (p_α) for the considered virtual motion is computed as the corresponding von Mises motor scalar product [14],

$$\delta W^{(\alpha)} = \underline{F}^{(\alpha)} \circ \delta \bar{k}^{(\alpha)} = \boldsymbol{F}^{(\alpha)} \cdot \delta \boldsymbol{v}^{\mathrm{K}_\alpha} + \boldsymbol{M}^{\mathrm{K}_\alpha} \cdot \delta \boldsymbol{w}^{(\alpha)} \tag{9.30}$$

Note that the von Mises motor scalar product is invariant of the choice of the particular selection of the reference point K_α.

The virtual power equation for particle (p_α) expressing equilibrium reads,

$$\delta W^{(\alpha)} = 0 \tag{9.31}$$

With Eq. (9.23) in Cartesian form, Eq. (9.31) becomes

$$\sum_{c \in C_\alpha} \left(f_i^{(\alpha)c} \delta v_i^{(\alpha)} + \left(m_i^{(a)c} + \varepsilon_{ijk} \left(x_j^c - x_j^\alpha \right) f_k^{(\alpha)c} \right) \delta w_i^{(\alpha)} \right) = 0 \tag{9.32}$$

Obviously, a particle assembly inside an (REV) is in equilibrium, if each sub-particle is in equilibrium. Then Eq. (9.32) holds for all grains in B, thus equilibrium for the assembly is expressed by

$$\sum_{\alpha \in B} \sum_{c \in C_\alpha} \left(f_i^{(\alpha)c} \delta v_i^{(\alpha)} + \left(m_i^{(a)c} + \varepsilon_{ijk} \left(x_j^c - x_j^\alpha \right) f_k^{(\alpha)c} \right) \delta w_i^{(\alpha)} \right) = 0 \tag{9.33}$$

The double sum over C_α and B can be split into two separate sums, one over I and one over E, respectively. In addition we observe that for any two grains (p_α) and (p_β) in contact at point P_c we have from Newton's 3rd law that,

$$\begin{aligned} f_i^{(\alpha,\beta)c} &\equiv f_i^{(\alpha)c} = -f_i^{(\beta)c} \quad \Rightarrow \quad f_i^{(\alpha,\beta)c} = -f_i^{(\beta,\alpha)c} \\ m_i^{(\alpha,\beta)c} &\equiv m_i^{(\alpha)c} = -m_i^{(\beta)c} \quad \Rightarrow \quad m_i^{(\alpha,\beta)c} = -m_i^{(\beta,\alpha)c} \end{aligned} \tag{9.34}$$

Thus from Eq. (9.33) we get an equivalent expression that can be written as a virtual power equation,

$$\delta W^{(\mathrm{int})} = \delta W^{(ext)} \tag{9.35}$$

where

$$\delta W^{(\mathrm{int})} = \sum_{c\in I}\left(f_i^{(\alpha,\beta)c}\left(\delta v_i^{(\alpha)c} - \delta v_i^{(\beta)c}\right) + m_i^{(\alpha,\beta)c}\left(\delta w_i^{(\alpha)} - \delta w_i^{(\beta)}\right)\right) \tag{9.36}$$

$$\delta W^{(ext)} = \sum_{e\in E}\left(f_i^{(e)}\delta v_i^{(e)} + m_i^{(e)}\delta w_i^{(e)}\right) \tag{9.37}$$

9.2.3 Equilibrium Conditions for Compatible Virtual Kinematics

In order to evaluate the above expressions, Eqs. (9.35) to (9.37), we set the virtual spin of a particle to be a linear function, and the virtual velocity of the centre of the particle to be a bi-linear form of the coordinates of the position of its centre,

$$\begin{aligned} \delta w_i^{(\alpha)} &= \alpha_i + \beta_{ij}x_j^\alpha \\ \delta w_i^{(\beta)} &= \alpha_i + \beta_{ij}x_j^\beta \\ &\dots \end{aligned} \tag{9.38}$$

$$\begin{aligned} \delta v_i^{(\alpha)} &= a_i + b_{ij}x_j^\alpha + c_{ijk}x_j^\alpha x_k^\alpha \\ \delta v_i^{(\beta)} &= a_i + b_{ij}x_j^\beta + c_{ijk}x_j^\beta x_k^\beta \\ &\dots \end{aligned} \tag{9.39}$$

where α_i, β_{ij} and a_i, b_{ij}, c_{ijk} are arbitrary coefficients. This bilinear expansion of the virtual velocity field has been used in the past by Chang & Liao [15] and was adopted later by Bardet & Vardoulakis [13]. Its justification, however, is supplied here as a consistency requirement by the consideration of the transport law for the particle velocity, Eq. (9.4).

If we introduce the particular realizations, Eqs. (9.38) and (9.39) in Eqs. (9.6) and (9.7) we get,

$$\delta w_i^{(\beta,\alpha)c} = \beta_{ij}\left(x_j^\beta - x_j^\alpha\right) \tag{9.40}$$

$$\begin{aligned} \delta v_i^{(\beta,\alpha)c} = {} & b_{ij}\left(x_j^\beta - x_j^\alpha\right) + c_{ijk}\left(x_j^\beta x_k^\beta - x_j^\alpha x_k^\alpha\right) - \varepsilon_{ijk}\alpha_j\left(x_k^\beta - x_k^\alpha\right) \\ & + \varepsilon_{ijk}\beta_{jl}\left(\left(x_k^c - x_k^\beta\right)x_l^\beta - \left(x_k^c - x_k^\alpha\right)x_l^\alpha\right) \end{aligned} \tag{9.41}$$

Similarly,

$$\begin{aligned}\delta v_i^e &= \delta v_i^{(\alpha)} + \varepsilon_{ijk}\delta w_j^{(\alpha)}\left(x_k^e - x_k^{\alpha e}\right)\\ &= a_i + b_{ij}x_j^{\alpha e} + c_{ijk}x_j^{\alpha e}x_k^{\alpha e} + \varepsilon_{ijk}\alpha_j\left(x_k^e - x_k^{\alpha e}\right) + \varepsilon_{ijk}\beta_{jl}x_l^{\alpha e}\left(x_k^e - x_k^{\alpha e}\right)\end{aligned} \tag{9.42}$$

where $x_i^{\alpha e}$ is the position of the centre of particle p_α, where contact e takes place.

From Eqs. (9.36) and (9.41) to (9.42) we get the following expressions for the virtual power of internal and external actions in and on the considered (REV),

$$\begin{aligned}\delta W^{(\text{int})} &= b_{ij}\sum_{c\in I} f_i^{(\alpha,\beta)c}\left(x_j^\beta - x_j^\alpha\right)\\ &+ c_{ijk}\sum_{c\in I} f_i^{(\alpha,\beta)c}\left(x_j^\beta x_k^\beta - x_j^\alpha x_k^\alpha\right)\\ &- \alpha_j\sum_{c\in I} f_i^{(\alpha,\beta)c}\varepsilon_{ijk}\left(x_k^\beta - x_k^\alpha\right)\\ &+ \beta_{ji}\sum_{c\in I}\left(\varepsilon_{ijk}f_i^{(\alpha,\beta)c}\left(x_l^\beta\left(x_k^c - x_k^\beta\right) - x_l^\alpha\left(x_k^c - x_k^\alpha\right)\right) + m_j^{(\alpha,\beta)c}\left(x_l^\beta - x_l^\alpha\right)\right)\end{aligned} \tag{9.43}$$

and

$$\begin{aligned}\delta W^{(ext)} &= a_i\sum_{e\in E} f_i^e + b_{ij}\sum_{e\in E} f_i^e x_j^{\alpha e}\\ &+ c_{ijk}\sum_{e\in E} f_i^e x_j^{\alpha e}x_k^{\alpha e}\\ &+ \alpha_j\sum_{e\in E}\left(m_j^e + \varepsilon_{ijk}f_i^e\left(x_k^e - x_k^{\alpha e}\right)\right)\\ &+ \beta_{ji}\sum_{c\in I}\left(m_j^e + \varepsilon_{ijk}f_i^e\left(x_k^e - x_k^{\alpha e}\right)\right)x_l^{\alpha e}\end{aligned} \tag{9.44}$$

The virtual power Eq. (9.35), with Eqs. (9.43) and (9.44), applies for arbitrary choice of the coefficients $a_i, b_{ij}, c_{ijk},\ \alpha_i, \beta_{ij}$. Thus, by independent variation of these coefficients we get the following set of algebraic equations:

$$\sum_{e\in E} f_i^e = 0 \tag{9.45}$$

$$\sum_{c\in I}\left(x_j^\beta - x_j^\alpha\right)f_i^{(\alpha,\beta)c} = \sum_{e\in E} x_j^{\alpha e}f_i^e \tag{9.46}$$

$$\sum_{c\in I} f_i^{(\alpha,\beta)c}\left(x_j^\beta x_k^\beta - x_j^\alpha x_k^\alpha\right) = \sum_{e\in E} f_i^e x_j^{\alpha e}x_k^{\alpha e} \tag{9.47}$$

$$\sum_{c \in I} \varepsilon_{ijk} \left(x_j^{\beta} - x_j^{\alpha} \right) f_k^{(\alpha,\beta)c} = - \sum_{e \in E} m_i^{(e,\alpha)} \tag{9.48}$$

$$\sum_{c \in I} \left(\varepsilon_{ikl} f_l^{(\alpha,\beta)c} \left(x_j^{\beta} \left(x_k^c - x_k^{\beta} \right) - x_j^{\alpha} \left(x_k^c - x_k^{\alpha} \right) \right) + m_i^{(\alpha,\beta)c} \left(x_j^{\beta} - x_j^{\alpha} \right) \right) = \sum_{e \in E} m_i^{(e,\alpha)} x_j^{\alpha e} \tag{9.49}$$

where

$$m_i^{(e,\alpha)} = m_i^e + \varepsilon_{ijk} \left(x_j^e - x_j^{\alpha e} \right) f_k^e \tag{9.50}$$

is the moment that results by transporting the external contact force and couple from point P_e on particle p_{α} to its centre K_{α}.

Equation (9.45) is expressing the equilibrium of external forces that are applied to the whole assembly of particles in the considered (REV). From Eq. (9.46) we get

$$\sum_{c \in I} \varepsilon_{ijk} \left(x_j^{\beta} - x_j^{\alpha} \right) f_k^{(\alpha,\beta)c} = \sum_{e \in E} \varepsilon_{ijk} x_j^{\alpha e} f_k^e \tag{9.51}$$

and with that Eq. (9.48) transforms into

$$\sum_{e \in E} \left(m_i^{(e,\alpha)} + \varepsilon_{ijk} x_j^{\alpha e} f_k^e \right) = 0 \tag{9.52}$$

or to the moment equilibrium equation for all external actions on the considered (REV),

$$\begin{aligned} &\sum_{e \in E} \left(m_i^e + \varepsilon_{ijk} \left(x_j^e - x_j^{\alpha e} \right) f_k^e + \varepsilon_{ijk} x_j^{\alpha e} f_k^e \right) = 0 \Rightarrow \\ &\sum_{e \in E} \left(m_i^e + \varepsilon_{ijk} x_j^e f_k^e \right) = 0 \end{aligned} \tag{9.53}$$

If we combine Eqs. (9.47) and (9.49) we obtain,

$$\sum_{c \in I} m_i^{(\alpha,\beta)c} \left(x_j^{\beta} - x_j^{\alpha} \right) = \sum_{e \in E} \left(m_i^e + \varepsilon_{inm} x_n^e f_m^e \right) x_j^{\alpha e} - \sum_{c \in I} \varepsilon_{ikl} f_l^{(\alpha,\beta)c} \left(x_j^{\beta} - x_j^{\alpha} \right) x_k^c \tag{9.54}$$

We summarize below the set of equations that we derived by applying the virtual work equation on an (REV) of particles that are in a state of static equilibrium under the action of external forces and couples,

$$\sum_{e \in E} f_i^e = 0 \tag{9.55}$$

$$\sum_{e\in E}\left(m_i^e + \varepsilon_{ijk}x_j^e f_k^e\right) = 0 \tag{9.56}$$

$$\sum_{c\in I}\left(x_j^\beta - x_j^\alpha\right)f_i^{(\alpha,\beta)c} = \sum_{e\in E} x_j^{\alpha e} f_i^e \tag{9.57}$$

$$\sum_{c\in I}\left(x_j^\beta x_k^\beta - x_j^\alpha x_k^\alpha\right)f_i^{(\alpha,\beta)c} = \sum_{e\in E} f_i^e x_j^{\alpha e} x_k^{\alpha e} \tag{9.58}$$

$$\sum_{c\in I}\left(x_j^\beta - x_j^\alpha\right)\left(m_i^{(\alpha,\beta)c} + \varepsilon_{ikl}x_k^c f_l^{(\alpha,\beta)c}\right) = \sum_{e\in E}\left(m_i^e + \varepsilon_{inm}x_n^e f_m^e\right)x_j^{\alpha e} \tag{9.59}$$

9.3 The Micromechanical Definition of Stress and Couple Stress

We consider a strategy for a transition from the discrete medium to the continuum. This is by far not a unique procedure, thus having always the character of a working hypothesis. The mathematical limitations of such strategies are discussed in detail by Froiio et al. [12].

For the computation of a mean value of the stress within the (REV) we follow a standard procedure [16]: The analysis starts from the stress equilibrium equations that apply for the continuum. We consider a small volume V of the continuum that in the discrete is occupied by the (REV) and we assume the existence of a stress field that satisfies the equilibrium equations on the considered volume and on its boundary. We express the equilibrium, Eq. (4.33), in Cartesian coordinates, we multiply these equations with x_k, integrate over V, apply Gauss' theorem and use Eq. (4.34),

$$\int_{V_{REV}} (\partial_i\sigma_{ij} + f_j)x_k dV = 0 \quad \Rightarrow \quad \int_{V_{REV}} \sigma_{kj} dV = \int_{\partial V_{REV}} x_k t_j dS + \int_{V_{REV}} x_k f_j dV = 0 \tag{9.60}$$

If the considered volume is a sphere with radius R_ε, then the surface integral on the r.h.s. of Eq. (9.60) is of $O\left(R_\varepsilon^2\right)$, whereas the volume integral is of $O\left(R_\varepsilon^3\right)$. Thus, as it was done already in the discrete medium analysis, the effect of volume forces will be neglected,

$$\int_{V_{REV}} \sigma_{kj} dV \approx \int_{\partial V_{REV}} t_j x_k dS \quad (R_\varepsilon \to 0) \tag{9.61}$$

We observe that the quantity

$$\hat{\sigma}_{ij} = \frac{1}{V_{REV}} \int\limits_{V_{REV}} \sigma_{ij} dV \tag{9.62}$$

is by definition the volume-averaged stress.

From Eqs. (9.61) and (9.62) we get,

$$\hat{\sigma}_{ij} \approx \frac{1}{V_{REV}} \int\limits_{\partial V_{REV}} t_j x_k dS \tag{9.63}$$

Note that according to some authors the above outlined stress averaging procedure can be traced to a reference of Love [17] on the work of Chree [18].

We juxtapose now Eqs. (9.61) and (9.57):

$$\int\limits_{\partial V_{REV}} x_k t_j dS = \int\limits_{V_{REV}} \sigma_{kj} dV \quad \leftrightarrow \quad \sum_{e \in E} x_k^{\alpha e} f_j^{e} = \sum_{c \in I} \left(x_k^{\beta} - x_k^{\alpha} \right) f_j^{(\alpha,\beta)c} \tag{9.64}$$

This analogy between the continuum and the discrete medium suggests a formula for the computation of the mean stress, that is evaluated by using micromechanical information inside the (REV),

$$\hat{\sigma}_{ij} \approx \frac{1}{V_{REV}} \sum_{c \in I} (x_i^{\beta} - x_i^{\alpha}) f_j^{(\alpha,\beta)c} \tag{9.65}$$

Equation (9.65) is a celebrated formula in granular Mechanics and Physics, that according to Fortin et al. [19] was first applied for the definition of the mean stress tensor in a granular medium in 1966 by Weber [20] and has been advocated since for this purpose by many authors; cf. [21, 22].

Similarly we assume the existence of a couple-stress field that satisfies the equilibrium equations for a Cosserat continuum, Eqs. (4.35) and (4.36). From the corresponding equilibrium equations written in Cartesian form we derive,

$$\int\limits_{V_{REV}} \mu_{kj} dV = \int\limits_{\partial V_{REV}} m_j x_k dS + \int\limits_{V_{REV}} \varepsilon_{imj} \sigma_{im} x_k dV \tag{9.66}$$

We remark that if surface couples and couple stresses are zero, then Eq. (9.66) reduces to a condition that implies symmetry of stress tensor. In general however, with

$$
\begin{aligned}
0 &= \int_{V_{REV}} (\partial_i \sigma_{ij}) x_k x_l dV = \int_{V_{REV}} \partial_i \left(\sigma_{ij} x_k x_l\right) dV - \int_{V_{REV}} \sigma_{ij} \partial_i (x_k x_l) dV \Rightarrow \\
&\int_{V_{REV}} \left(\sigma_{kj} x_l + \sigma_{lj} x_k\right) dV = \int_{\partial V_{REV}} t_j x_k x_l dS
\end{aligned}
\tag{9.67}
$$

the last integral on the r.h.s. of Eq. (9.66) becomes,

$$
\begin{aligned}
\int_{V_{REV}} \varepsilon_{imj} \sigma_{im} x_k dV &= \int_{V_{REV}} \varepsilon_{imj} \sigma_{[im]} x_k dV = - \int_{V_{REV}} \varepsilon_{imj} \sigma_{[mi]} x_k dV \\
&= - \int_{V_{REV}} \varepsilon_{imj} \sigma_{mi} x_k dV = \int_{V_{REV}} \varepsilon_{imj} \sigma_{ki} x_m dV - \int_{\partial V_{REV}} \varepsilon_{imj} t_i x_k x_m dS
\end{aligned}
\tag{9.68}
$$

and with that

$$
\int_{V_{REV}} \mu_{kj} dV = \int_{\partial V_{REV}} m_j x_k dS + \int_{V_{REV}} \varepsilon_{imj} \sigma_{ki} x_m dV - \int_{\partial V_{REV}} \varepsilon_{imj} t_i x_k x_m dS \tag{9.69}
$$

or

$$
\int_{V_{REV}} \left(\mu_{kj} + \varepsilon_{jmi} x_m \sigma_{ki}\right) dV = \int_{\partial V_{REV}} \left(m_j + \varepsilon_{jmi} x_m t_i\right) x_k dS \tag{9.70}
$$

The tensor

$$
\hat{\mu}_{kj} = \frac{1}{V_{REV}} \int_{V_{REV}} \left(\mu_{kj} + \varepsilon_{jmi} x_m \sigma_{ki}\right) dV \tag{9.71}
$$

is called here the "mean transported couple stress", computed over the volume V_{REV}. Equations (9.70), (9.71) and (9.59) juxtaposed suggest that,

$$
\hat{\mu}_{ij} \approx \frac{1}{V_{REV}} \sum_{c \in I} \left(x_j^{\beta} - x_j^{\alpha}\right) \left(m_i^{(\alpha,\beta)c} + \varepsilon_{ikl} x_k^c f_l^{(\alpha,\beta)c}\right) \tag{9.72}
$$

This formula was originally introduced by Chang & Liao [15] and later by Bardet & Vardoulakis [6] and Tordesillas & Walsh [23]. This definition differs from the one that was proposed by Oda & Iwashita [24] that was proposed in turn in complete formal analogy to Eq. (9.65),

$$\tilde{\mu}_{kj} \approx \frac{1}{V_{REV}} \sum_{c \in I} \left(x_k^{\beta} - x_k^{\alpha} \right) m_j^{(\alpha,\beta)c} \tag{9.73}$$

The existence of the two definitions, Eqs. (9.72) and (9.73), explains the controversy in relation to the statements that couple-stresses in granular media are: a) only due to contact couples [24], an assumption that would support definition (9.73), or b) that they are also generated in part also by the contact forces, as in definition Eq. (9.72). We emphasize here that the definition of $\hat{\mu}_{ij}$, Eq. (9.72), is derived following a procedure that is in complete analogy to the derivation of Weber's formula for stress, Eq. (9.62). This is not possible for the couple stress $\tilde{\mu}_{ij}$. We will demonstrate below that $\hat{\mu}_{ij}$ as a statically meaningful measure of the transported to the grain-contacts couple stress, is also meaningful from an energetic point of view.

9.4 Intergranular Dissipation

Rigid granular media are dissipative media in the sense that all energy supplied to them by the external actions is dissipated. As stated by Cole and Peters [25] "... *the relationship between the contact motions and resisting forces define the micro-scale properties of the medium*..." In that sense central in our approach here are energy dissipation considerations.

As is shown in Fig. 9.6, the contact of two homothetically rotating grains will involve strong contact sliding and weak contact rolling, whereas the contact of two antithetically rotating grains will involve strong contact rolling and weak contact sliding. A basic hidden assumption made in earlier studies was that almost all energy dissipation in granular media is localized at sliding contacts [26]. In general, however, energy dissipation due to rolling cannot be excluded due to micro slip and friction at the contact interface. In this context we like to refer the reader directly to the discussion offered on the subject by Tordesillas and Walsh [23]. This point of view is appreciated as a fact by many investigators, since in a number of recent DEM simulations, energy dissipation is admitted to rolling contacts as well [27–29].

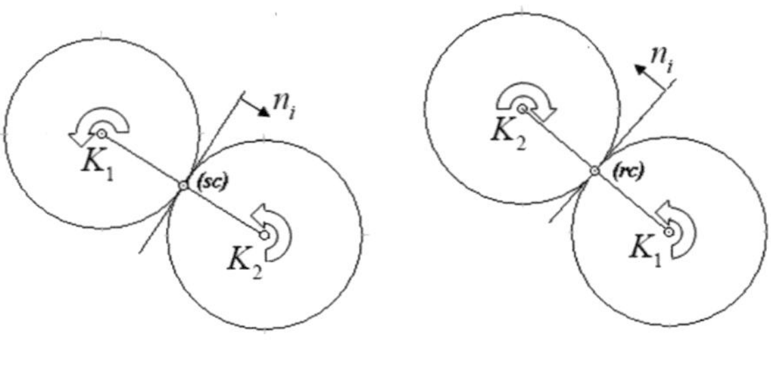

Fig. 9.6 Two grain circuit with sliding contact and rolling contact respectively

9.4.1 Grain Contact Energetics

We consider two homothetically rotating grains and for simplicity we set them to be of equal radius R_g with a strong sliding contact, as seen in Fig. 9.7. The branch vector that connects the centers of the two grains is,

$$(K_1K_2)_i = 2\ell_i \quad ; \quad \ell_i = R_g n_i \ , \ n_k n_k = 1 \tag{9.74}$$

The velocities of the centers of the grains (1) and (2) are denoted by $v_i^{(1)}$ and $v_i^{(2)}$, and the grains are rotating homothetically with angular velocities $w_k^{(1)}$ and $w_k^{(2)}$, respectively. At the midpoint P_c of the centre line (K_1K_2) the velocities of the grains are

$$\begin{aligned} v_i^{(1)c} &= v_i^{(1)} + \varepsilon_{ilk} w_l^{(1)} \ell_k \\ v_i^{(2)c} &= v_i^{(2)} - \varepsilon_{ilk} w_l^{(2)} \ell_k \end{aligned} \tag{9.75}$$

Thus the relative rotation and relative velocity of grain (2) with respect to grain (1) at the contact point are,

$$w_i^{(2,1)c} = w_i^{(2)} - w_i^{(1)} \tag{9.76}$$

$$v_i^{(2,1)c} = v_i^{(2)c} - v_i^{(1)c} = v_i^{(2)} - v_i^{(1)} - \varepsilon_{ijk}\left(w_j^{(2)} + w_j^{(1)}\right)\ell_k \tag{9.77}$$

Since the two grains are in contact, we assume that they interact with contact forces and contact couples. Let $f_i^{(1,2)c}$ and $m_i^{(1,2)c}$ be the force and the couple acted on grain (1) by grain (2); their reactions are the contact force $f_i^{(2,1)c}$ and the contact couple $m_i^{(2,1)c}$, acting on grain (2). These force- and couple pairs satisfy Newton's 3rd law,

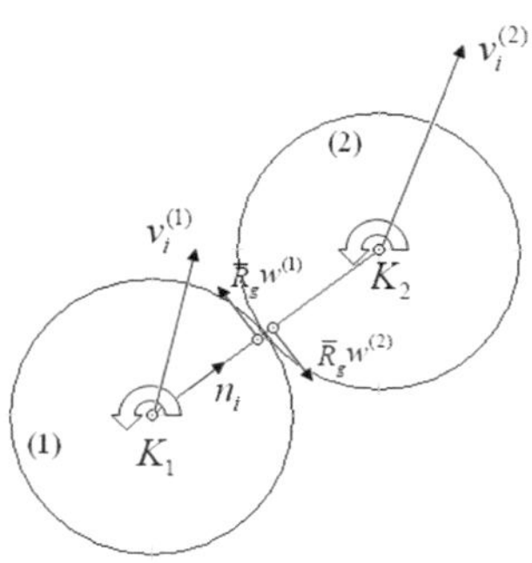

Fig. 9.7 Two-grain circuit: kinematic embedment

$$\begin{aligned} f_i^{(1,2)c} &\equiv f_i^{(1)c} = -f_i^{(2)c} \quad \Rightarrow \quad f_i^{(1,2)c} = -f_i^{(2,1)c} \\ m_i^{(1,2)c} &\equiv m_i^{(1)c} = -m_i^{(2)c} \quad \Rightarrow \quad m_i^{(1,2)c} = -m_i^{(2,1)c} \end{aligned} \tag{9.78}$$

The interface at the contact of the two grains is identified as an *intergranular surface*. This is a continuum material band of vanishing thickness, whose boundaries share the motion of the two adjacent faces of the contact. In the terminology of Tribology this interface is called the "third body". As stated by Godet [30]: "Interfaces, or third bodies can be defined in a material sense, as a zone which exhibits a marked change in composition from that of the rubbing specimens or in a kinematic sense, as the thickness across which the difference in velocity between solids is accommodated".

On the faces of this infinitesimal slip the reactions of the intergranular forces are acting. On the face of the intergranular surface that touches point $P_c^{(2)}$ and has the outer unit normal n_i, the force $f_i^{(1,2)c}$ is acting. On the opposite face of the intergranular surface that touches point $P_c^{(1)}$ and has the outer unit normal $-n_i$, the force $f_i^{(2,1)}$ is acting. The rate of work per unit volume, done by these forces at the considered contact due to sliding, is (Fig. 9.8),

$$P^{(ns)} = \frac{1}{V}\left(f_i^{(1,2)} v_i^{(2c)} + f_i^{(2,1)} v_i^{(1c)}\right) = \frac{1}{V} f_i^{(1,2)}\left(v_i^{(2c)} - v_i^{(1c)}\right) = \frac{1}{V} f_i^{(1,2)} v_i^{(2,1)c} \tag{9.79}$$

Similarly the rate of work of contact couples at the considered contact due to (weak) rolling is (Fig. 9.9),

$$P^{(nr)} = \frac{1}{V}\left(m_i^{(1,2)c} w_i^{(2)} + m_i^{(2,1)c} w_i^{(2)}\right) = \frac{1}{V} m_i^{(1,2)c}\left(w_i^{(2)} - w_i^{(1)}\right) = \frac{1}{V} m_i^{(1,2)c} w_i^{(2,1)} \tag{9.80}$$

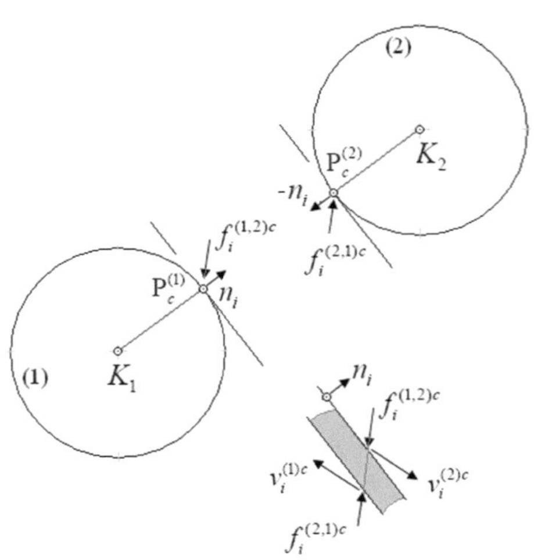

Fig. 9.8 Two-grain circuit: strong sliding contact

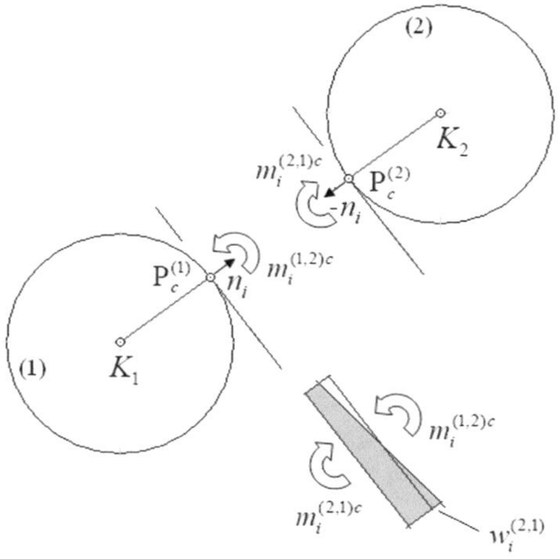

Fig. 9.9 Two-grain circuit: weak rolling sliding contact

The total power of actions at grain contact is the sum of the contributions due to sliding and rolling,

$$P^{(n)} = P^{(ns)} + P^{(nr)} \tag{9.81}$$

9.4.2 Continuum Embedment

We assume that particle spin and particle (centre) velocity are embedded into continuous fields, such that

$$w_i^{(1)} = w_i, \quad w_i^{(2)} \approx w_i + 2\ell_m \partial_m w_i \tag{9.82}$$

and

$$v_i^{(1)} = v_i, \quad v_i^{(2)} \approx v_i + 2\ell_m \partial_m v_i \tag{9.83}$$

With

$$w_i^{(2,1)c} = w_i^{(2)} - w_i^{(1)} = 2\ell_m \partial_m w_i \tag{9.84}$$

and

$$v_i^{(2,1)} = v_i^{(2)} - v_i^{(1)} - \varepsilon_{ijk}\left(w_j^{(2)} + w_j^{(1)}\right)\ell_k = 2\left(\partial_k v_i - \varepsilon_{ijk} w_j + \varepsilon_{ikj} \ell_m \partial_m w_j\right)\ell_k \tag{9.85}$$

we observe that the relative spin and velocity can be expressed in terms of the related Cosserat-continuum deformation measures; c.f. Equations (3.138) and (3.139). With this notation Eqs. (9.84) and (9.85) become,

$$w_i^{(2,1)} = 2\ell_m K_{im} \tag{9.86}$$

and

$$v_i^{(2,1)c} = 2(\Gamma_{ki} + \varepsilon_{ikl} K_{ml} \ell_m)\ell_k \tag{9.87}$$

On the other hand, we assume that the force $f_i^{(1,2)}$ and the couple $m_i^{(1,2)}$ are generated by a stress field and a couple stress field, respectively, which are evaluated in turn at the centre of the considered grain. Thus

$$\frac{f_i^{(1,2)}}{S} = \sigma_{ki}(-n_k)\ , \quad \frac{f_i^{(2,1)}}{S} = \sigma_{ki}(+n_k) \tag{9.88}$$

where S is a reference surface area, and with that

$$P^{(ns)} = \left(\frac{S}{V}\right)\left(\sigma_{ki}(-n_k)v_i^{(1c)} + \sigma_{ki} n_k v_i^{(2c)}\right) = \frac{1}{a}\sigma_{ki} n_k \left(v_i^{(2c)} - v_i^{(1c)}\right) = \frac{1}{a}\sigma_{ki} n_k v_i^{(2,1)} \tag{9.89}$$

The specific surface

$$\frac{S}{V} = \frac{1}{a} \tag{9.90}$$

is a free parameter that will be determined below with the requirement that the power of contact actions at grain scale relates to the stress power in the underlying Cosserat continuum. Due to Eq. (9.87) from Eq. (9.89) we get

$$P^{(ns)} = \frac{1}{a}\sigma_{ki} n_k 2(\Gamma_{ni} + \varepsilon_{inl} \mathrm{K}_{ml} \ell_m)\ell_n = \frac{2R_g}{a}(\sigma_{ki} n_k)((\Gamma_{ni} n_n) + \varepsilon_{inl}(\mathrm{K}_{ml} n_m)\ell_n) \tag{9.91}$$

Let

$$\lambda = \frac{2R_g}{a} \tag{9.92}$$

With the notation

$$t_i^{(n)} = \sigma_{ki} n_k \tag{9.93}$$

and

$$\Gamma_i^{(n)} = \Gamma_{ki} n_k; \quad \mathrm{K}_i^{(n)} = \mathrm{K}_{ki} n_k \tag{9.94}$$

we observe that the vector compound

$$\bar{E}^{(n)} = \begin{pmatrix} \mathbf{K}^{(n)} \\ \boldsymbol{\Gamma}^{(n)} + \boldsymbol{l} \times \mathbf{K}^{(n)} \end{pmatrix} \tag{9.95}$$

is a von Mises deformation-vector motor for all possible contact points of grain (1), with unit normal on the contact plane $\boldsymbol{n}$. The "moment" part of the kinematic motor (9.95) works on the stress vector, since

$$P^{(ns)} = \lambda \boldsymbol{t}^{(n)} \cdot \left(\Gamma^{(n)} + l \times \mathrm{K}^{(n)}\right) \tag{9.96}$$

In view of Eq. (9.71) we postulate the couple-stress field $\mu_{ij}^{(c)}$, called the contact transported couple stress, such that,

$$\mu_{ij}^{(c)} = \mu_{ij} + \varepsilon_{jlk} \ell_l \sigma_{ik} \tag{9.97}$$

and we define the moment vectors,

$$m_i^{(n)} = \mu_{ki} n_k \tag{9.98}$$

and

$$m_i^{(cn)} = \mu_{ki}^{(c)} n_k = (\mu_{ki} + \varepsilon_{ilm} \ell_l \sigma_{km}) n_k = m_i^{(n)} + \varepsilon_{ilm} \ell_l t_m^{(n)} \tag{9.99}$$

We observe that the compound

$$\underline{T}^{(n)} = \begin{pmatrix} \boldsymbol{t}^{(n)} \\ \boldsymbol{m}^{(n)} + \boldsymbol{l} \times \boldsymbol{t}^{(n)} \end{pmatrix} \tag{9.100}$$

is a von Misses action-vector motor for all possible contact points of grain (1).

With these remarks we compute the von Mises scalar product of the two motors and claim that this is the total power of actions at a grain contact,

$$\frac{1}{\lambda} P^{(n)} = \underline{T}^{(n)} \circ \bar{E}^{(n)} = t^{(n)} \cdot \left(\boldsymbol{\Gamma}^{(n)} + \boldsymbol{l} \times \mathbf{K}^{(n)}\right) + \left(\boldsymbol{m}^{(n)} + \boldsymbol{l} \times \boldsymbol{t}^{(n)}\right) \cdot \mathbf{K}^{(n)} \tag{9.101}$$

We showed already through Eq. (9.96) that the first term on the r.h.s of Eq. (9.101) reflects the work done by the forces due to (strong) sliding, Eq. (9.96). We will prove now that the 2nd term on the r.h.s. of Eq. (9.101) corresponds to the work done by the couples at the considered contact due to (weak) rolling.

Indeed, if we define after Eq. (9.80) that

$$\begin{aligned} P^{(nr)} &= \frac{1}{V}\left(m_i^{(1,2)c} w_i^{(1)} + m_i^{(2,1)c} w_i^{(2)}\right) = \left(\frac{S}{V}\right)\left(\mu_{ki}^{(c)}(-n_k)w_i^{(1)} + \mu_{ki}^{(c)} n_k w_i^{(2)}\right) \\ &= \frac{1}{a}\mu_{ki}^{(c)} n_k \left(w_i^{(2)} - w_i^{(1)}\right) = \frac{1}{a}\mu_{ki}^{(c)} n_k w_i^{(2,1)} = \frac{2R_g}{a}\mu_{ki}^{(c)} n_k \ell_m \partial_m w_i \end{aligned} \tag{9.102}$$

we get

$$\frac{1}{\lambda} P^{(nr)} = \left(\boldsymbol{m}^{(n)} + \boldsymbol{l} \times \boldsymbol{t}^{(n)}\right) \cdot \mathbf{K}^{(n)} \tag{9.103}$$

and with that we recover Eq. (9.81).

We return to Eq. (9.101) and observe that mixed terms cancel out, leading finally to,

$$\frac{1}{\lambda} P^{(n)} = \boldsymbol{t}^{(n)} \cdot \boldsymbol{\Gamma}^{(n)} + \boldsymbol{m}^{(n)} \cdot \mathbf{K}^{(n)} \tag{9.104}$$

or in Cartesian form,

$$\frac{1}{\lambda} P^{(n)} = (\sigma_{ki}\Gamma_{li} + \mu_{ki}\mathrm{K}_{li}) n_k n_l \tag{9.105}$$

9.4.3 Fabric Averaging

Let $\boldsymbol{n}$ be the unit normal that characterizes an intergranular contact plane at contact point P_c, as this was discussed in the previous sections. We select all such contact normal vectors and transfer them parallelly to the centre of the unit sphere. This mapping defines a point P'_c on the unit sphere. The distribution of these points on the unit sphere defines in turn an essential property of the fabric of the contact-planes network. The simplest assumption is that the probability distribution of the unit contact normal vectors is uniform. This assumption is rather crude as far as granular media are concerned, and for realistic modeling cconsiderations it should be replaced by suitable anisotropic probability distributions [31, 32].

We observe that the corresponding Cartesian coordinates of the position vector $\overrightarrow{\mathrm{OP}'_c} = \boldsymbol{n}$ on the unit sphere are

$$n_1 = \sin\theta\,\cos\phi; \quad n_2 = \sin\theta\,\sin\phi; \quad n_3 = \cos\theta \tag{9.106}$$

where $r = 1$, θ and ϕ are the polar, spherical coordinates of point P'_c. In case of isotropy, it can be easily shown that following identities hold [33]:

$$
\begin{aligned}
&\langle n_i \rangle = 0\\
&\langle n_i n_j \rangle = \frac{1}{4\pi}\int_0^{2\pi}\int_0^{\pi} n_i n_j \sin\theta d\theta d\phi = \frac{1}{3}\delta_{ij}\\
&\langle n_i n_j n_k \rangle = 0\\
&\langle n_i n_j n_k n_l \rangle = \frac{1}{3\cdot 5}\left(\delta_{ij}\delta_{kl} + \delta_{ik}\delta_{jl} + \delta_{il}\delta_{jk}\right) = \frac{1}{15}\delta_{ijkl}\\
&\langle n_i n_j n_k n_l n_m \rangle = 0\\
&\langle n_i n_j n_k n_l n_m n_n \rangle = \frac{1}{3\cdot 5\cdot 7}\left(\delta_{in}\delta_{jklm} + \delta_{jn}\delta_{klmi} + \delta_{kn}\delta_{lmij} + \delta_{\ln}\delta_{mijk} + \delta_{mn}\delta_{ijkl}\right) = \frac{1}{105}\delta_{ijklmn}\\
&\ldots
\end{aligned}
\tag{9.107}
$$

We return to the expression for the power of contact forces and contact couples at intergranular contact, Eq. (9.105), and compute its average. For

$$
\lambda = \frac{2R_g}{a} = 3 \quad \Rightarrow \quad a = \frac{V}{S} = \frac{2}{3}R_g \tag{9.108}
$$

and

$$
P^{(n)} = 3(\sigma_{ki}\Gamma_{li} + \mu_{ki}\mathrm{K}_{li})n_k n_l \tag{9.109}
$$

we get finally,

$$
\left\langle P^{(n)} \right\rangle = P = \sigma_{ij}\Gamma_{ij} + \mu_{ij}\mathrm{K}_{ij} \tag{9.110}
$$

where P is the power of internal actions in the Cosserat continuum, cf. Equation (4.3).

Thus the particular choice of micromechanical variables at the level of intergranular contact has allowed us to recover the stress power of the Cosserat continuum as the isotropic average value of the work done by contact forces and contact couples at the third body at grain contacts.

9.5 Stress- and Couple-Stress Invariants for Isotropic Fabric

We define the scalar

$$
t^{(n)} = t_i n_i = \sigma_{ki} n_k n_i \tag{9.111}
$$

that is the normal component of the stress vector acting on a contact plane with unit normal vector $\boldsymbol{n}$.

The vector

$$t_i^{(t)} = t_i - t^{(n)} n_i = \sigma_{ki} n_k - \sigma_{kl} n_k n_l n_i \tag{9.112}$$

is the corresponding shear stress vector, with

$$t_i^{(t)} n_i = 0 \tag{9.113}$$

By introducing the decomposition of the stress tensor in spherical and deviatoric part,

$$\sigma_{ij} = s_{ij} + \frac{1}{3}\sigma_{kk}\delta_{ij} \pm \quad , \quad s_{kk} = 0 \tag{9.114}$$

we get

$$t^{(n)} = \left(s_{ik} + \frac{1}{3}\sigma_{mm}\delta_{ik} \right) n_i n_k \tag{9.115}$$

$$t_j^{(t)} = s_{ik}\left(\delta_{jk} - n_j n_k\right) n_i \tag{9.116}$$

By averaging over all probable normal-contact directions in the considered (REV), from Eqs. (9.115), (9.116) and (9.107) we get the following statistical stress invariants:

(a) The mean normal traction on contact

$$\left\langle t^{(n)} \right\rangle = p = \frac{1}{3}\sigma_{kk} \tag{9.117}$$

(b) The mean of the square of the magnitude of the shear traction on contact,

$$\left\langle t_i^{(t)} t_i^{(t)} \right\rangle = \frac{1}{3}\left(\frac{4}{5} s_{kp} s_{kp} - \frac{1}{5} s_{kp} s_{pk} \right) \tag{9.118}$$

If we apply Eq. (9.118) for the symmetric part of the stress tensor, then the average of the square of the shear stress magnitude is related to the usual shearing stress intensity [34],

$$\tau_{mean} = \sqrt{\frac{1}{3}\left(\frac{4}{5} s_{(kp)} s_{(kp)} - \frac{1}{5} s_{(kp)} s_{(pk)} \right)} = \sqrt{\frac{1}{5} s_{(kp)} s_{(pk)}} \tag{9.119}$$

or

$$\tau_{mean} = \sqrt{\frac{2}{5}}\mathrm{T} \tag{9.120}$$

where

$$\mathrm{T} = \sqrt{\frac{1}{2}s_{kp}s_{kp}} \tag{9.121}$$

We recall that in case of a Boltzmann continuum the shearing stress intensity T differs but little from the maximum shear stress,

$$0.87\max(\tau_{\max}) \leq \mathrm{T} \leq \min(\tau_{\max}) \tag{9.122}$$

and is used extensively in the formulation of maximum shear stress criteria.

We repeat the above procedure for the contact couple-stress tensor, $\mu_{ij}^{(c)}$, that was defined above through Eq. (9.97)

$$\mu_{ij}^{(c)} = \mu_{ij} + R_g \varepsilon_{jlk} \sigma_{ik} n_l \tag{9.123}$$

Let

$$m_j^{(cn)} = \mu_{ij}^{(c)} n_i = \mu_{ij} n_i + R_g \varepsilon_{jlk} \sigma_{ik} n_i n_l \tag{9.124}$$

be the normal component and

$$m^{(cn)} = m_j^{(cn)} n_j = \mu_{ij} n_i n_j + R_g \varepsilon_{jlk} \sigma_{ik} n_i n_l n_j \tag{9.125}$$

The 1st statistical moment of the contact couple-stress tensor $\mu_{ij}^{(c)}$ is a measure for the mean contact torsion,

$$\mu_T = \left\langle m^{(cn)} \right\rangle = \mu_{ij} \langle n_i n_j \rangle + R_g \varepsilon_{jlk} \sigma_{ik} \langle n_i n_l n_j \rangle = \frac{1}{3}\mu_{kk} \tag{9.126}$$

We observe that the mean torsion is transported unaltered from the stress field. This allows us to decompose the contact couple-stress tensor into a spherical and a deviatoric part as,

$$\mu_{ij}^{(c)} = m_{ij}^{(c)} + \mu_T \delta_{ij} \tag{9.127}$$

where

$$m_{ij}^{(c)} = \mu_{ij} + R_g \varepsilon_{jlk} \sigma_{ik} n_l - \mu_T \delta_{ij} = m_{ij} + R_g \varepsilon_{jlk} \sigma_{ik} n_l \tag{9.128}$$

With

$$m_j^{(ct)} = m_{ik}^{(c)}\left(\delta_{jk} - n_j n_k\right) n_i \tag{9.129}$$

we can compute the 2nd statistical moment of the deviatoric contact couple-stress tensor, $\mu_{ij}^{(c)}$,

$$m_{mean}^{(c)} = \sqrt{\left\langle m_s^{(ct)} m_s^{(ct)} \right\rangle} \tag{9.130}$$

In analogy to Eq. (9.118) we have

$$m_{mean}^{(c)} = \frac{1}{3}\left(\frac{4}{5} m_{kp}^{(c)} m_{kp}^{(c)} - \frac{1}{5} m_{kp}^{(c)} m_{pk}^{(c)}\right) \tag{9.131}$$

or

$$\mathrm{M}^{(c)} = \sqrt{\frac{5}{2} m_{mean}^{(c)}} = \sqrt{\frac{4}{6} m_{kp}^{(c)} m_{kp}^{(c)} - \frac{1}{6} m_{kp}^{(c)} m_{pk}^{(c)}} \tag{9.132}$$

that we call the rolling-contact couple-stress intensity.

Using the definition of the contact couple-stress, Eq. (9.123), we can express $m_{mean}^{(c)}$ in terms of the couple-stress and stress deviators,

$$m_{mean}^{(c)} = \sqrt{\left\langle m_s^{(ct)} m_s^{(ct)} \right\rangle} = \sqrt{m_{mean}^2 + R_g^2 \tau_{mean}^2} \tag{9.133}$$

where

$$m_{mean} = \sqrt{\frac{1}{3}\left(\frac{4}{5} m_{kp} m_{kp} - \frac{1}{5} m_{kp} m_{pk}\right)} \tag{9.134}$$

We define the corresponding deviatoric couple-stress intensity

$$\mathrm{M} = \sqrt{\frac{5}{2} m_{mean}} = \sqrt{\frac{4}{6} m_{kp} m_{kp} - \frac{1}{6} m_{kp} m_{pk}} \tag{9.135}$$

and with this notation Eq. (9.132) becomes

$$\mathrm{M}^{(c)} = \sqrt{\mathrm{M} + R_g^2 \mathrm{T}} \tag{9.136}$$

or explicitly,

$$\mathrm{M}^{(c)} = \sqrt{\left(\frac{4}{6}m_{kp}m_{kp} - \frac{1}{6}m_{kp}m_{pk}\right) + R_g^2\left(\frac{4}{6}s_{kp}s_{kp} - \frac{1}{6}s_{kp}s_{pk}\right)} \tag{9.137}$$

The above introduced stress- and couple-stress invariants, can be used in the formulation of plasticity based constitutive equations for granular media, to express the ability of the material to provide resistance to external actions due to irreversible interparticle slip, torsion and rolling, respectively. In case that someone wishes to generalize plasticity models that incorporate in their formulation the effect of the 3rd invariant as well, then one could consider the computation of appropriate 3rd order "moments" of the deviators of the stress tensor σ_{ij} and the contact couple stress tensor $\mu_{ij}^{(c)}$.

At this point we remark that the present analysis is based on the use of the transport laws for velocity and force of rigid-body mechanics in the realm of Cosserat continuum approximation of the mechanics of granular assemblies. This analysis has resulted in a drastic modification of our previous Cosserat plasticity models for granular materials [34, 35]. Based on the early work of Besdo [36] on Cosserat plasticity for ductile materials, in the aforementioned work no distinction was made between sliding and rolling contacts. To this end an ad hoc definition of a compound stress was introduced, that had the form,

$$\tilde{\sigma}_{ij} = \sigma_{ij} + \frac{1}{R^*}\varepsilon_{ijk}\mu_{jl}n_l \tag{9.138}$$

The definition of this extra stress was inspired in turn from Schaefer's [37] analogy between Cosserat continuum theory and beam theory. On that basis stress invariants were computed that resulted in a modified shearing stress intensity of the form,

$$\tilde{T} = \sqrt{\mathrm{T}^2 + \frac{1}{R^{*2}}\mathrm{M}^2} \tag{9.139}$$

that was used in turn in the formulation of single-yield surface plasticity theories.

References

1. Satake, M. (1968). *Some considerations on the mechanics of granular materials*. In E. Kröner (Ed.), *Mechanics of generalized continua* (pp. 156–159). Berlin: Springer.
2. Günther, W. (1958). Zur Statik und Kinematik des Cosseratschen Kontinuums. *Abhandlungen der Braunschweigische Wissenschaftliche Gesellschaft, 10,* 195–213.
3. Kröner, E. (1992). *Plastizität der Versetzungen*. In A. Sommerfeld (Ed.), *Mechanik der deformierbaren Medien* (pp. 310–376). Verlag Harri Deutsch: Frankfurt.

4. Rivier, N. (2006). Extended constraints, arches and soft modes in granular materials. *Journal of Non-Crystalline Solids, 352,* 4505–4508.
5. Oda, M., & Kazama, H. (1998). Microstructure of shear bands and its relation to the mechanisms of dilatancy and failure of dense granular soils. *Géotechnique, 48,* 465–481.
6. Satake, M. (1998). *Finite difference approach to the shear band formation from viewpoint of particle column buckling*. In *Thirteenth Southeast Asian Geotechnical Conference*. Taipei, Taiwan: ROC.
7. Dietrich, T. (1976). *Der psammische Stoff als Modell des mechanischen Sandes*. Universität Karlsruhe.
8. Vardoulakis, I. (1989). Shear banding and liquefaction in granular materials on the basis of a Cosserat continuum theory. *Ingenieur Archiv, 59,* 106–113.
9. Radjai, F., et al. (1998). Bimodal character of stress transmission in granular packings. *Physical Review Letters, 80,* 61–64.
10. Staron, L., Vilotte, J. P., & Radjai, F. (2001). *Friction and mobilization of contacts in granular numerical avalanches*. In *Powders & grains 2001*. Sendai, Japan: Swets & Zeitlinger.
11. Tordesillas, A. (2007). Force chain buckling, unjamming transitions and shear banding in dense granular assemblies. *Philosophical Magazine, 87,* 4987–5016.
12. Froiio, F., Tomassetti, G., & Vardoulakis, I. (2006). Mechanics of granular materials: The discrete and the continuum descriptions juxtaposed. *International Journal of Solids and Structures, 43*(25–26), 7684–7720.
13. Bardet, J. P., & Vardoulakis, I. (2001). The asymmetry of stress in granular media. *International Journal of Solids and Structures, 38,* 353–367.
14. Brand, L. (1940). *Vector and tensor analysis*. John Wiley.
15. Chang, C. S., & Liao, L. L. (1990). Constitutive relation for a particulate medium with the effect of particle rotation. *International Journal of Solids and Structures, 26*(4), 437–453.
16. Landau, L. D., & Lifshitz, E. M. (1959). *Theory of elasticity*. Course of theoretical physics. Oxford: Pergamon Press.
17. Love, A. E. H. (1927). *A treatise of the mathematical theory of elasticity*. Cambridge: Cambridge University Press.
18. Chree, C. (1892). Changes in the dimension of elastic bodies due to given systems of forces. *Transactions of the Cambridge Philosophical Society, 15,* 313–337.
19. Fortin, J., Millet, O., & de Saxcé, G. (2002). Mean stress in a granular medium in dynamics. *Mechanics Research Communications, 29*(4), 235–240.
20. Weber, J. (1966). Recherche concernant les contraintes intergranulaires dans les milieux pulverulents. *Bulletin de Liaison Ponts et Chausses, 20,* 1–20.
21. Christoffersen, J., Mehrabadi, M. M., & Nemat-Nasser, S. (1981). A micromechanical description of granular material behavior. *ASME Journal of Applied Mechanics, 48*(2), 339–344.
22. Rothenburg, L., & Selvadurai. A. P. S. (1981). *A micromechanical definition of the Cauchy stress tensor for particular media*. In *International Symposium on the Mechanical Behavior of Structured Media*. Ottawa: Elsevier.
23. Tordesillas, A., & Walsh, D. C. S. (2002). Incorporating rolling resistance and contact anisotropy in micromechanical models of granular media. *Powder Technology, 124,* 106–111.
24. Oda, M., & Iwashita, K. (2000). Study on couple stress and shear band development in granular media based on numerical simulation analyses. *International Journal of Engineering Science, 38,* 1713–1740.
25. Cole, D. M., & Peters, J. F. (2007). A physically based approach to granular media mechanics: Grain-scale experiments, initial results and implications to numerical modeling. *Granular Matter, 9*(5), 309–321.
26. Mehrabadi, M. M., & Cowin, S. C. (1978). Initial planar deformation of dilatant granular materials. *Journal of the Mechanics and Physics of Solids, 26,* 269–284.
27. Jiang, M. J., Yu, H. S., & Harris, D. (2005). A novel discrete model for granular material incorporating rolling resistance. *Computers and Geotechnics, 32*(5), 340–357.

28. Alonso-Marroquin, F., et al. (2006). Effect of rolling on dissipation in fault gouges. *Physical Review E, 74,* 031306.
29. Estrada, N., Taboada, A., & Radjai, F. (2008). Shear strength and force transmission in granular media with rolling resistance. *Physical Review E, 78,* 021301.
30. Godet, M. (1990). Third-bodies in tribology. *Wear, 136,* 29–45.
31. Hamel, G. (1921). Elementare Mechanik. *Zeitschrift für Angewandte Mathematik und Mechanik, 1*(3), 219–223.
32. Satake, M. (1982). *Fabric tensor in granular materials.* In *IUTAM Conference on Deformation and Failure of Granular Materials.* Delft: Balkema.
33. Kanatani, K. (1979). A micropolar continuum theory for flow of granular materials. *International Journal of Engineering Science, 17,* 419–432.
34. Vardoulakis, I., & Sulem, J. (1995). *Bifurcation analysis in geomechanics.* Blackie Academic & Professional.
35. Mühlhaus, H.-B., & Vardoulakis, I. (1987). The thickness of shear bands in granular materials. *Géotechnique, 37,* 271–283.
36. Besdo, D. (1974). Ein Beitrag zur nichtlinearen Theorie des Cosserat-Kontinuums. *Acta Mechanica, 20,* 105–131.
37. Schaefer, H. (1967). Analysis der Motorfelder in Cosserat-Kontinuum. *Zeitschrift für Angewandte Mathematik und Mechanik, 47,* 319–332.

Printed in the United States
By Bookmasters